Technological Innovation Networks

A volume in
Contemporary Perspectives on Technological Innovation, Management and Policy
Bing Ran, *Series Editor*

Technological Innovation Networks

Collaboration and Partnership

edited by

Bing Ran

Pennsylvania State University at Harrisburg

IAP

INFORMATION AGE PUBLISHING, INC.

Charlotte, NC • www.infoagepub.com

Library of Congress Cataloging-in-Publication Data

A CIP record for this book is available from the Library of Congress
http://www.loc.gov

ISBN: 978-1-68123-858-6 (Paperback)
978-1-68123-859-3 (Hardcover)
978-1-68123-860-9 (ebook)

Printed and bound by CPI Group (UK) Ltd, Croydon, CR0 4YY

CONTENTS

PREFACE

The central theme of this book series is to explore contemporary perspectives on managing technological innovations and related strategic policy issues. I had two objectives when editing this book: one, to be open to all potential topics that I believe need attention within the broad theme of the management of technology and innovations; and two, to promote an interdisciplinary scholarship and dialogue on the management of innovation and technological change in a global context from strategic, managerial, behavioral, and policy perspectives.

The third volume of this book series concentrates on "Technological Innovation Networks: Collaboration and Partnership"—a theme resonating with scholars and practitioners that innovation requires a network of partners to collaborate. Authors from around the world contribute to this volume by approaching this theme from many different perspectives: an institutional understanding of international R&D networks, a stakeholder centrality potential in innovation networks, the intersection between intellectual structure and M&A, the rejections of the technological opportunities due to lock-in, the policy–practice paradox of technological innovations, Japan's national innovation strategy, immigrant entrepreneurs in patents and performance, the impact of university research parks on technology transfer, a historical narrative of cotton technology in China, and the innovative online or blended education in terms of motivation and reality. These researches have made significant attempts to address the important questions of how technological innovation touched on many aspects of our networked social life, and thus I hope readers who are interested in

Technological Innovation Networks, pages vii–viii

learning the most contemporary perspectives on the technological innovation will be impressed, enriched, and intrigued by their analyses in each chapter. As the editor, I hope readers of the volume can enjoy these chapters by their global nature, the practicality orientation, the critical perspective, and the new theories and practices embedded in the selected research.

Let the journey begin here.

—**Bing Ran**
Pennsylvania State University at Harrisburg

CHAPTER 1

THE INSTITUTIONAL UNDERSTANDING OF TECHNOLOGICAL INNOVATION

Considerations on International R&D Networks

Federica Brunetta, Luca Giustiniano, and Paolo Boccardelli
LUISS Guido Carli University

ABSTRACT

We aim at observing the impact of institutional frameworks on firm performance paying specific attention to mechanisms through which institutions and legal frameworks shape the behaviors, actions and relations in R&D networks. We observed international networks and their R&D location choices. In doing so, we tried to understand how institutional forces rationalize the environment in which the network is embedded and eventually leverage network relations. We focus on actors that are embedded in several institutional environments, where they are offered the possibility to seize different institu-

Technological Innovation Networks, pages 1–17

> tional opportunities present in other parts of the network, as diverse institutional spheres can represent windows in which frameworks tend to constrain the actions of firms to a lesser degree. We address the following research questions: How are institutional frameworks impacting on the formation of international networks and R&D location choices? Can international networks be the result of strategies of jurisdiction shopping? Under which circumstances do firms engage in institutional entrepreneurship strategies?

Decisions about location can be driven by the characteristics of the institutional environment. In the field of innovation, locating R&D units affects the conditions under which firms interact with the environment. Innovation occurs frequently under interfirm umbrellas (Pennings & Harianto, 1992a) and high levels of innovation interdependency among firms are found in many industries. Linkages become of strategic importance (Pennings, 1981; Pennings & Harianto, 1992a, 1992b; Perks & Jeffrey, 2006) due to the increasing complexity of technology bases and levels of technological sophistication (Freeman, 1991), cross-fertilization of scientific disciplines and fields of technology (Hagedoorn, 1993), R&D intensity and technologic uncertainty (DeBresson & Amesse, 1991), rapidity of technological development, and broad distribution of research breakthroughs (Powell, Koput, & Smith-Doerr, 1996; Shan, Walker, & Kogut, 1994). The extant literature has focused on the impact of collaborations on the innovative performance of focal firms, analyzing the importance of networking to access technology that is alien to the firm (Pennings & Harianto, 1992a, 1992b), to share innovation risks and costs, and to accelerate the rate of new product development to a certain extent (Deeds & Hill, 1996).

Considering the centrality of location for the actual effects of institutions, scholars are claiming a deeper understanding of the institutional environment in which these R&D networks are embedded (Powell, White, Koput, & Owen-Smith, 2005). In order to foster understanding of managing technological innovations and related policy and strategy issues we adopt the institution-based view (IBV). The focus on IBV has been recently addressed by several strategy scholars (Ahuja & Yayavaram, 2011; Peng, Sun, Pinkham, & Chen, 2009; Peng, Wang, & Jiang, 2008) as the third prominent perspective in strategic management (the first two being the industry-based and resource-based views).

Literature recognizes that institutions are both the medium for, and the result of, social action: They enable and constrain what firms and other agents wish to accomplish (Morgan, 2004). Indeed, institutions, defined by North (1990, p. 3) as the "humanly devised rules of the game in a society" as well as the social structures that create, embody, and enforce those rules, do "directly determine what arrows a firm has in its quiver as it struggles to formulate and implement strategy, and to create a competitive advantage" (Ingram & Silverman, 2002, p. 20).

Innovation literature often refers to national innovation systems (NISs), related to a country's R&D, education, financial systems, as well as structure and capacity of the system as a whole (Morgan, 2004). NISs are usually the framework under which literature explains the cognitive capabilities of firms within a nation—including multinationals—molding the way they look at the environment, the way they act, how they assess opportunities and risks. NISs usually allow and foster certain courses of actions, while not always determining specific behaviors. Given the strong emphasis on both formal and informal—normative—institutions, it is important to take into account the IBV and the impact that institutions have on firm strategy and technological management, as well as their organizational and economic outcomes.

The roots of IBV are to be found in the new institutionalism. According to the new institutional perspective, once disparate organizations in the same line of business are structured into an actual field, isomorphism emerges as a consequence of the re-alignment between organizational and environmental characteristics (DiMaggio & Powell, 1983; Meyer & Rowan, 1977). Thus, the interconnections between organizations and the institutional environment (DiMaggio & Powell, 1983) can be defined as institutional embeddedness. The subject has also been at the center of a recent vivid point/counterpoint between Greenwood, Hinings, and Whetten (2014) and Meyer and Höllerer (2014) published in the *Journal of Management Studies.*

Each country possesses an environment consisting of a unique policy and regulatory framework that is part of the industry's investment–innovation cycle and links and covers the critical elements of the value chain. The literature recognizes that this holds true even for strongly globalized industries and is demonstrated by the fact that national patterns of development do not converge or evolve around shared "world norms" (Morgan, 2004, p. 13). For instance, countries exhibit very different patterns of technological specialization (Morgan, 2004; Patel & Pavitt, 1994).

Driven by large multinational companies willing to exploit national advantages and different levels of specialization, the trend towards the internationalization of R&D has been clear and constant since the 1980s. Large multinational companies have first internationalized their innovation activities (Gerybadze & Reger, 1999; Nonaka & Takeuchi, 1995; Patel & Pavitt, 1994). Grandstrand, Hakanson, and Sjolander (1993) have recognized that government regulations have played an important role in inducing firms to set up locally adaptive R&D, as drivers of decentralizing R&D. It is therefore clear that different government policies have a substantial impact on key aspects of the innovation performance (Jommi & Paruzzolo, 2003). Within countries, forms of specialization and advantage are additionally based on subnational formations, such as clusters, into regional innovation systems. We analyze the impact of institution on R&D location choices and the role of institution in providing incentives to innovation.

By using the IBV perspective, underlining its connection to the fields of strategy and innovation management, we analyze the role of institutions in managing technological innovations and related policy and strategy issues. Therefore, we specifically formulate research propositions for international R&D networks. As the topic provides fertile ground for academic and managerial implications, we will finally reflect upon the implications both from the point of view of the firms and from the policymaker point of view. We address the following research questions: *How are institutional frameworks impacting on the formation of international networks and R&D location choices? Can international networks be the result of strategies of jurisdiction shopping? Under which circumstances do firms engage in institutional entrepreneurship strategies?*

With an emphasis on broad institutional environments and societal considerations (DiMaggio & Powell, 1983; Meyer & Rowan, 1977; Meyer & Scott, 1983; Scott, 1995), institutional theory has been widely used for studying the adoption of organizational practices among organizations, as well as the interaction between institutions and organizations that encourages normative and mimetic isomorphism behaviors over institutional constraints. When the environment is highly regulated, firms tend towards homogenization through processes of isomorphic change (Hawley, 1968). Firms converge towards procedural conformity (Scott, 1987) that constrains organizations in the way they perform specific tasks or procedures following specific requirements that are imposed by institutions.

THE ROLE OF R&D NETWORKS FOR INNOVATION

In 1934, Schumpeter proposed a first definition of innovation as a new combination of productive resources. Since then, the concept of innovation has changed and is now seen as an interactive process involving relationships among different actors (Kline & Rosenberg, 1986), a process involving exchanging knowledge (Patel and Pavitt, 1994) or the interaction of learning and exchange genereating an innovative system or cluster (Edquist, 1997).

In this context, firms opt for cooperative efforts in order to reduce uncertainties associated with novel products or markets rather than developing internal capabilities necessary for success, so that the establishment of different types of cooperative interorganizational relationships (Oliver, 1990) is common. Indeed, in the strategy field, a logic of dyadic relationships is often prevalent, and those who study the business community are most likely studying alliances, interorganizational relationships, coalitions, or collaborative agreements. Strategic alliances are one of the ways that firms respond to the complexity of the environment in which they operate, and they are defined as "voluntary arrangements between firms involving exchange, sharing, or co-development of products, technologies or services" (Gulati, 1998, p. 293); the benefits of sharing resources such as skills, knowledge, and information about competitors, customers, suppliers, and markets have long been argued by (Hamel, Doz, & Prahalad, 1989), and the literature shows that linkages enable firms to achieve economies of scale and scope, in addition to learning new, or enhancing existing capabilities (Kogut, 1988). Moreover, strategic alliances provide access to the technological knowledge of partners and leverage complementary assets (Teece, 1992). Literature shows general consensus on the benefits of strategic technological alliances on innovative output, especially on patent and new product development. Following Podolny and Page (1998), nevertheless, different forms of cooperation agreements among two or more actors engaged in collaboration and lacking a legitimate authority can themselves be thought of as a network form of organization including joint ventures, strategic alliances, business groups, franchises, research consortia, and outsourcing agreements.

Critical innovation resources frequently reside in the network and not in the firm alone (Afuah, 2000; Arora & Gambardella, 1990; Freeman 1991; Powell et al., 1996). Stemming from the definition by Laumann, Galaskiewicz, and Marsden (1978) of networks as "a set of nodes (people, organizations) linked by a set of social relationships of a specified type" (p. 458), extending the network concept to organizations relaxing Mitchell's (1969) definition based on network of individuals, networks have been broadly studied by scholars who have observed nodes, ties, and outcomes of different typologies. Over time, they have been defined as dynamic networks, in which major components can be assembled or reassembled to meet changing and competitive conditions (Miles & Snow 1986); strategic networks, when the goal is long-term, purposeful arrangements among distinct but related for-profit organizations that allow those firms within them to gain or sustain competitive advantage (Gulati, Nohria, & Zaheer, 2000; Jarillo, 1988); interfirm networks, as modes of organizing economic activities through interfirm cooperation and cooperation (Grandori & Soda 1995); and finally, with respect to innovation, as networks of innovators in which certain types of interorganizational linkages are appropriate for technological transactions and developing innovation (DeBresson & Amesse, 1991).

In high-growth, technology intensive industries, collaboration networks can be seen as organizational devices needed to coordinate heterogeneous learning processes of agents possessing diverse skills, competencies, access to information, and assets (Arora & Gambardella, 1994; Barley, Freeman, & Hybels, 1992; Orsenigo, Pammolli, & Riccaboni, 2001; Pisano, 1991; Powell et al., 1996; Walker, Kogut, & Shan, 1997).

Innovating firms working together can be thought of as research networks (DeBresson & Amesse, 1991). This definition is close to that of Doz,

Olk, and Ring (2000, p. 240) of an R&D consortium, seen as a legal entity established by two or more organizations that pool resources and share decision making for cooperative research and development activities.

Nevertheless, the focus of this chapter is on the networks intended as loci of innovation. As more complex systems of cooperation, interaction, or collaboration among different actors (driven by the increasing complexity and multidisciplinarity of research) make very difficult to identify a single innovator, Arorea and Gambardella (1990) have underlined the need to move towards a view of the network as the ideal locus for innovation. In the same context, Giustiniano and D'Alise (2015) explained some of these effects with reference to the resource-dependence theory.

From this perspective, Powell et al. (1996) in their work on learning, have identified networks serving as loci for innovation in those networks providing timely access to knowledge and resources that are otherwise unavailable and stimulating internal expertise and learning capabilities, especially in industries in which complex knowledge bases expand rapidly. Gulati (1999) also underlines the role of networks as the alternative access to resources that are not readily available through market exchanges. The position of firms influences firm behavior and outcomes as networks offer greater intellectual depth, opportunity scanning, competence enhancement and value-added solutions, or a worldwide reach (Powell et al., 1996; Powell, Koput, Smith-Doerr, & Owen-Smith, 1999; Walker et al., 1997).

With reference to innovation and technology development, networks offer two substantial benefits, as identified by Ahuja (2000), that can be traced to resource sharing and combination of different knowledge and skills and to the access to knowledge spillovers; the former enabling the transfer of know-how and physical assets, the latter facilitating transfer of information.

In this light, network structure enables three different mechanisms, according to direct and indirect ties, and to connections between partners (Ahuja, 2000). Specifically, the number of direct ties a firm maintains can positively affect its innovative output by providing benefits of knowledge sharing. In fact, firms collaborating to develop a technology are likely to share resultant knowledge (Berg, Duncan, & Friedman, 1982), complementarities in skills (Arora & Gambardella, 1990), and scale, with a resultant level of resources which is proportionally higher than that of a single firm (Ahuja, 2000). Indirect ties, which are less costly to maintain, are likely to foster the diffusion of knowledge through knowledge spillovers in activities that are often information-intense and therefore enhance the innovation output, although their impact decreases when the presence of direct ties increases (Ahuja, 2000).

INSTITUTIONAL FRAMEWORKS AND R&D LOCATION CHOICES

We refer to institutional embeddedness as the interconnections between organizations and the institutional environment (DiMaggio & Powell, 1983). We argue that when actors are embedded in different institutional environments, they are offered the possibility to seize different institutional opportunities present in other parts of the network. Diverse institutional spheres can represent windows in which frameworks tend to constrain the actions of firms to a lesser degree.

We aim to analyze the mechanisms that shape the behaviors, actions, and relations among actors (Zukin & DiMaggio, 1994). Institutional forces rationalizing the environment in which the network is embedded may leverage network relations or break social ties and the consequent benefits of embeddedness (Uzzi, 1997). Moreover, innovation policies are usually focused on the creation of a particular environment and certain attitudes that enable coordination between the agents directed to achieving innovation (Sweeney, 1995). Policies create "cultures of cooperation" that are fostered and nurtured by links that are not only determined by economic regions but which may be driven by cultural, political, or normative rules (Cooke, Uranga, & Etxebarria, 1997). Trust building is essential for firms involved in innovation processes to understand informal and formal protocols.

Research processes, because of their nature, do not have clear schedules as the outcomes emerge as a result of changing conditions (Bechky, 2006) and different applicable regulations. Specifically, we formulate a research proposition that can be analyzed in settings characterized by institutional forces that significantly constrain actors subjected to increasing specialization and systemic complexity, such as the pharmaceutical or the energy industry.

Organizations often establish network relations to meet necessary or regulatory requirements due to mandates from higher authorities that enhance cooperation mechanisms that otherwise might not occur voluntarily in the light of necessity as opposed to other drivers like asymmetry (the exercise of control over another organization), reciprocity (collaboration, cooperation), efficiency (the effort to increase the input–output ratio), stability (response to environmental uncertainties) or legitimacy (reputation, image, prestige) (Oliver, 1990).

An R&D collaboration can be defined as the practice through which a firm establishes a relationship with an external organization with the purpose of improving the performance of its R&D processes (Chiesa & Toletti, 2004). The competitive mechanism associated with stringent regulatory environments and non-competitive dynamics on the final markets tend to promote innovation (Gambardella, Orsenigo & Pammolli, 2001).

The neoinstitutional model states that organizational survival is determined by the degree of alignment with the institutional environment (DiMaggio & Powell, 1983; Meyer & Rowan, 1977; Scott, 1995).

Each country possesses an environment consisting of a unique policy and regulatory framework that is part of the industry's investment–innovation cycle and links and covers the critical elements of the value chain. One of the primary requirements of investors in innovation is to operate in a predictable, stable environment; each region evolves along different trajectories through combinations of political, cultural, and economic forces (Cooke et al., 2007). It is therefore clear that different regional and governmental policies have a substantial impact on key aspects of the innovation performance (Jommi & Paruzzolo, 2007).

National and regional systems become an institutional repository of institutional routines, norms, and values through which actors may come to trust each other collectively (North, 1990, 1993; Scott, 1987). Institutions reduce uncertainty structuring daily life and providing guidance—not necessarily in an efficient way—to human actions. So, they include both formal rules, and informal constraints, composed by non-written codes that function as a framework for interpretation

Informal constraints are the initial type of institution and arise when information is limited and no formal rule is in place, reducing the cost of interaction. Relationships are guided by codes of conduct, norms of behavior and conventions. As they derive from socially transmitted information and are part of cultures and traditions (North, 1990), those constraints have different meanings according to the different social contexts and may lead to changes at the formal level. In the short term, culture defines the way in which individuals process and use the information, and therefore it shapes the genesis and the characteristics of informal constraints.

Conventions and norms of behavior are, for instance, connected to culture. In the long term, culture plays a relevant role in the evolution of informal constraints. What is culturally derived tends to be resilient to change at the formal level. Indeed, several market problems arise from the tension between formal and informal constraints. Formal constraints include what is forbidden but also the conditions under which certain activities may be carried out. They are made up of formal and written rules. Although formal rules do not always and immediately change the existing informal framework, they are often put in place to impel that change. More specifically, formal constraints include legal, political, and economic rules, following a precise hierarchy flowing from constitutions to regulations, bylaws, and contracts (North, 1990).

Different institutional settings are likely to give rise to distinctive conventions or forms of collective social order leading to the enhancement of different kinds of organizations and even, to some extent, rules of the regulation determining the rise of different network relations. This leads us to the formulation of the first research proposition:

> **Proposition 1:** *Different institutional frameworks influence the establishment of international networks and the structure of the network itself.*

Some authors (Casper & Matraves, 2003; Jommi & Paruzzolo, 2007) have highlighted that the regulatory framework is one of the factors influencing the R&D locational choices. The trend towards the internationalization of R&D has been clear and constant since the 1980s and is driven by large multinational companies that have first internationalized innovation activities (Gerybadze & Reger, 1999; Nonaka & Takeuchi, 1995; Patel & Pavitt, 1994). Grandstrand et al. (1993) have recognized that government regulations have played an important role in inducing firms to set up local adaptive R&D, as drivers of decentralization of R&D.

Network structure is influenced by external conditions, which refer to the knowledge base and technological conditions in the industry, as well as uncertainty and industry events (Ozman, 2009); that is, Orsenigo et al. (Orsenigo, Pammolli, Riccaboni, Bonaccoris, & Turchetti, 1998; Orsenigo et al., 2001) point out that the hierarchical structure of network of firms can be perceived as an adaptive response to the hierarchical structure of the knowledge base. In the dynamics of networks, actors interact under different social mechanisms and different behavioral rules (Ozman, 2009).

The institutional-based view (IBV) defines *rent of influence* (Ahuja & Yayavaram, 2011) as the extra profit that can be earned due to the fact that the institutional "rules of the game" are designed to favor an economic actor or group of actors (Ahuja & Yayavaram, 2011). Laws, regulations, informal norms, and social structures regulate and moderate the behavior of actors within an environment and render implementation and enforcement possible. In this context, actors can achieve extra profits according to their potential or actual power to influence the institutions. This type of situation arises if an actor has the possibility to influence the definition of rules, or if the rules are unclear, or finally, if the rule or the institution do not exist.

Firms can engage in different generic strategies to generate influence rents by managing how institutions work and eventually fail. Logically, firms can pursue two broad generic strategies, namely avoidance of institutions and manipulation of institutions, or eventually resist institutional pressures (Ahuja & Yayavaram, 2011). Within the generic avoidance strategies, five generic actions can be identified, namely delaying, substitution, defanging, jurisdiction shopping (also defined as forum shopping or jurisdictional arbitrage), and arbitragive morphing. At the generic manipulation strategies level, firms can engage in subversion, starvation, perception management, co-optation and capture, or institutional proliferation. Finally, firms may choose to resist institutional pressures through a buffering or a decoupling strategy.

While focusing on the different elements of the aforementioned strategies is not the purpose of this work (see Oliver, 1991, and Ahuja & Yayavaram, 2011, for an in-depth analysis), we do focus on the strategy of *jurisdiction shopping* (Ahuja & Yayavaram, 2011) and on the idea of *institutional entrepreneurship* (Di Maggio, 1988).

Using a jurisdiction shopping strategy (Ahuja and Yayavaram, 2011), companies avoid the control of the institution by selecting a geographic location in which to operate on the basis of a more favorable institutional framework. When actors engage in these kinds of strategies, competition among regulatory spheres may arise, leading countries and geographic areas to seek to attract firms by shaping the institutional regime to the firms' advantage, such as the use of "pioneer rights," or entry barriers protecting early movers into a market (Ahuja & Yayavaram, 2011).

Distinct differences exist in national regulation systems, and the literature concurs in indicating the regulatory framework as one of the factors influencing R&D location choices; indeed, the trend towards the internationalization of R&D has been clear and constant since the 1980s and is driven by large multinational companies that have first internationalized innovation activities (Grandstrand et al., 1993). Government regulations might then play an important role in inducing firms to set up local adaptive R&D, as drivers of decentralization of R&D (Grandstrand et al., 1993). In the light of strategies of jurisdiction shopping, firms may be influenced in the R&D location choices decision by regulation, institutional constraints and opportunities.

> **Proposition 2:** *Jurisdiction Shopping Strategies may impact on the R&D location decision.*

When firms do not engage in avoiding institutions, but intend to shape the institutional solutions, they may activate processes of *institutional entrepreneurship*. The institutional entrepreneur is someone who "has an interest in certain institutional solutions and leverages the resources to create new institutions or to transform existing ones" (Maguire, Hardy, & Lawrence, 2004, p. 657). Originally, the concept of an institutional entrepreneur was proposed by DiMaggio (1988), who stressed that new institutions arise when organized actors with sufficient resources see in them the opportunity to realize interests that they assess as being of great value.

The institutional entrepreneur must be able to change the existing rules associated with prevailing institutional logics and institutionalize alternative rules (Garud, Hardy, & Maguire, 2007). To be able to push through the change in fields whose organizations are favored by the existing institutional solutions, it is necessary that the institutional entrepreneur adopts strategies that allow for the formation and transformation of institutions, fields, rules, and standards (Lawrence, 1999). In this light, entrepreneurs may adopt three strategies: occupy positions characterized by broad legitimacy, connecting the various stakeholders; formulate new practices; or connect the new practices to stakeholders' routines (Maguire et al., 2004).

We argue that firms may decide to engage in institutional entrepreneurship strategies aimed at shaping ~~of~~ the existing institutional frameworks in order to influence the organization's ability to resist institutional pressures rather than engaging in avoidance strategies and jurisdiction shopping. This leads us to our third and last research proposition:

> **Proposition 3:** *The institutional framework influences the decision to engage in institutional entrepreneurship strategies.*

DISCUSSION AND CONCLUSIONS

Managers stress that international presence is often guided by R&D needs. For example, in a recent interview (Burns, 2013), Eugenio Aringhieri, CEO of Dompé (a leading Italian pharmaceutical company headquartered in Milan, active in developing innovative treatment solutions for rare, often orphan diseases) stated:

> **Q:** What do you hope to accomplish with the opening of a New York subsidiary? Where in NY will the office be located? Will the NY location be the first U.S. location for Dompé? Are there plans to open additional U.S. offices?
>
> **A:** Dompé is currently carrying out an internationalization plan aimed at reaching out new geographic areas. In this perspective, the opening of a subsidiary in the United States to directly follow the ongoing trials and strengthen its R&D network represents a precise strategic choice.

This evidence is anecdotal. In fact, in the pharmaceutical industry there are many differences in countries' regulations and their implementation, although all the regulation frameworks are led and guided by the same ethical principles (Brunetta, Boccardelli, & Lipparini, 2015; Brunetta & Peruffo, 2014; Capo, Brunetta, & Boccardelli, 2014). However, the appraisal reported shows that R&D location choices and institutions have an important impact on the innovative performance of R&D. These differences, indeed, appear to be limited to some aspects of the research process but to have a strong impact on the innovative performance.

Managerial and Policy Implications

Managerial implications can be drawn both from the point of view of the firms that lead the development phases and design the structure of the research network and from the policymaker point of view. One of the primary requirements when investing in innovation is to operate in an environment providing predictable, stable institutions protecting and offering incentives to ~~the~~ innovative activity.

As firms are subject to institutional scrutiny, we highlighted how they may opt for different strategies when responding to institutional pressures. Coping with the pressures, or even avoiding them, makes firms more responsive to given institutional schemas, but it is recognized that serious development of an IBV should also take into consideration limits to the effectiveness of such strategic behavior (Ahuja & Yayavaram, 2011).

Drawing on an example from the context of pharmaceutical development, policy must pursue different objectives at the same time, guaranteeing safe and effective products (health policy), promoting an efficient use of resources (optimal combination of innovation and competition), stimulating pharmaceutical industries, and controlling costs once the product reaches the market. The strong regulation framework involves the presence of two different phenomena: the importance of operating within institutional spheres where opportunities for firms may arise and the shaping of the existing institutional frameworks in order to influence the organization's ability to resist institutional pressures.

Directions for Future Research and Limitation of the Study

Our research highlights some issues still to be addressed. First, in formulating our proposition, we have mainly focused on national systems and country regulations and policies, while it is recognized that within the same national environment, different regions coexist and can reflect diverse institutional schemas, build up on the basis their proactivity, and possess different identities deriving from the competencies they possess, which, in turn, can create new "social capital," or capacity for networking practices (Cooke et al., 1997). Subgroups are critical mesolevels within the system that in turn is composed of a set of integrated substructures (Frank & Yasumoto, 1998); these entities are also likely to mediate between the dyadic level and the systemic coherence; this, indeed, can be an important dynamic in highly regulated contexts. Second, we overlooked an important element in the innovation fostering mechanisms, which is related to financial systems. Markets depend critically on the preservation of individual incentives to produce and trade. This is one of the main market problems (Ahuja & Yayavaram, 2011) that institutions solve by ensuring the right to enjoy the fruits of entrepreneurial efforts in order to keep innovators motivated. Financing mechanisms vary from country to country and are linked to national and regional idiosyncrasies and formal and informal institutions.

REFERENCES

Afuah A. (2000). How much do your co-opetitors' capabilities matter in the face of technological change? *Strategic Management Journal, 21*(3), 387–404.

Ahuja G. (2000). Collaboration networks, structural holes, and innovation: A longitudinal study. *Administrative Science Quarterly, 45*(3), 425–455.

Ahuja, G., & Yayavaram, S. (2011). Explaining influence rents: The case for an institutions based view of strategy. *Organization Science, 22*(6), 1631–1652.

Arora A., & Gambardella A. (1990). Complementarity and external linkages: The strategies of the large firms in biotechnology. *Journal of Industrial Economics,* 38(4), 361–379.

Arora, A., & Gambardella, A. (1994). Evaluating technological information and utilizing it: Scientific knowledge, technological capability, and external linkages in biotechnology. *Journal of Economic Behavior & Organization, 24*(1), 91.

Barley, S., Freeman, J., & Hybels, T. (1992). Strategic alliances in commercial biotechnology. In N. Norhia & R. Eccles (Eds.), *Networks and organizations: Structure, form and action* (pp. 311–347). Boston, MA: Harvard Business School Press.

Bechky, B. A. (2006). Gaffers, gofers, and grips: Role-based coordination in temporary organizations. *Organization Science, 17*(1), 3–21.

Berg, S., Duncan, J., & Friedman, P. (1982). *Joint venture strategies and corporate innovation.* Cambridge, MA: Oelgeschlager, Gunn & Hain.

Brunetta, F., Boccardelli, P., & Lipparini, A. (2015). Central positions and performance in the scientific community. Evidences from clinical research projects. *Journal of Business Research, 68*(5), 1074–1081.

Brunetta, F., & Peruffo, E. (2014). May parents inherit from heirs? Towards an understanding of the parent-spun-off relationship. *American Journal of Applied Sciences, 11*(6), 821.

Burns, M. (2013) Dompé carries out internationalization plan. *Drugs.com.* Retrieved from http://www.drugs.com/clinical_trials/domp-carries-out-internationalization-plan-16340.html

Capo, F., Brunetta, F., & Boccardelli, P. (2014). Innovative business models in the pharmaceutical industry: A case on exploiting value networks to stay competitive. *International Journal of Engineering Business Management, 23*(6), 1–11.

Casper, S., & Matraves, C. (2003). Institutional frameworks and innovation in the German and UK pharmaceutical industry. *Research Policy, 32*(10), 1865–1879.

Chiesa V., & Toletti G. (2004). Network of collaborations for innovation: The case of biotechnology. *Technology Analysis and Strategic Management, 16*(1), 73–96.

Cooke, P., Uranga, M. G., & Etxebarria, G. (1997). Regional innovation systems: Institutional and organisational dimensions. *Research Policy, 26*(4), 475–491.

DeBresson C, & Amesse F. (1991). Networks of innovators: A review and introduction to the issue. *Research Policy, 20*(5), 363–379.

Deeds, D. L., & Hill, C. W. L. (1996). Strategic alliances and the rate of new product development: An empirical study of entrepreneurial biotechnology firms. *Journal of Business Venturing, 11*(1), 41–55.

DiMaggio, P. (1988). Interest and agency in institutional theory. In L. Zucker (Ed.), *Institutional patterns and organizations: Culture and environment.* Cambridge, MA: Ballinger.

DiMaggio, P., & Powell, W. (1983). The iron cage revisited: Institutional isomorphism and collective rationality in organizational fields. *American Sociological Review, 48*(2), 147–160.

Doz, Y., Olk, P., & Ring, P. (2000). Formation processes of R&D consortia: Which path to take? Where does it lead? *Strategic Management Journal, 21*(3), 239–266.

Edquist, C. (1997). *Systems of innovation: Technologies, institutions, and organizations.* New York, NY: Psychology Press.

Frank, K. A., & Yasumoto, J. Y. (1998). Linking action to social structure within a system: Social capital within and between subgroups 1. *American Journal of Sociology, 104*(3), 642–686.

Freeman, C. (1991). Networks of innovators: A synthesis of research issues. *Research Policy, 20*(5), 499–514.

Gambardella, A., Orsenigo, L., & Pammolli, F. (2001). *Global competitiveness in pharmaceuticals: A European perspective* (Enterprise Paper N°1). Brussels, BE: European Commission.

Garud, R., Hardy, C., & Maguire, S. (2007). Institutional entrepreneurship as embedded agency: An introduction to the special issue. *Organization Studies, 28*(7), 957.

Gerybadze, A., & Reger, G. (1999). Globalization of R&D: recent changes in the management of innovation in transnational corporations. *Research Policy, 28*(2-3), 251–274.

Giustiniano, L., & D'Alise, C. (2015). Networks, clusters, and small worlds. *Journal of Organization Design, 4*(2), 48–53

Grandori A., & Soda G. (1995). Inter-firm networks: Antecedents, mechanisms and forms. *Organization Studies, 16*(2), 183–214

Grandstrand, O., Hakanson, L., & Sjolander, S. (1993). Internationalization of R&D—A survey of some recent research. *Research Policy, 22*(5-6), 413–430.

Greenwood, R., Hinings, C. R., & Whetten, D. (2014). Rethinking institutions and organizations. *Journal of Management Studies, 51*(7), 1206–1220.

Gulati R. (1998). Alliances and networks, *Strategic Management Journal, 19*(4), 293–317.

Gulati, R. (1999). Network location and learning: The influence of network resources and firm capabilities on alliance formation. *Strategic Management Journal, 20*(5), 397–420.

Gulati, R., Nohria, N., & Zaheer, A. (2000). Strategic networks. *Strategic Management Journal, 21*(3), 203–215.

Hagedoorn, J. (1993). Understanding the rationale of strategic technology partnering: interorganizational modes of cooperation and sectoral differences. *Strategic Management Journal, 14*(5), 371–386.

Hamel, G., Doz, Y., & Prahalad, C. (1989). Collaborate with your competitors and win. *Harvard Business Review, 67*(1), 133–139

Hawley A. (1968). Human ecology. In D. Sills (Ed.), *International encyclopaedia of social sciences* (pp. 328–337). New York, NY: Macmillan.

Ingram, P. L., & Silverman, B. S. (2002). *The new institutionalism in strategic management.* New York, NY: Elsevier.

Jarillo, J. C. (1988). On strategic networks. *Strategic Management Journal, 9*(1), 31–41.

Jommi, C., & Paruzzolo, S. (2007). Public administration and R&D localisation by pharmaceutical and biotech companies: A theoretical framework and the Italian case-study. *Health Policy 81*(1), 117–130.

Kline, J., & Rosenberg, N. (1986). An overview of innovation. In R. Landau, & N. Rosenberg (Eds.), *The positive sum strategy: Harnessing technology for economic growth* (pp. 275–306). Washington, DC: National Academic Press.

Kogut B. (1988). Joint ventures: Theoretical and empirical perspectives. *Strategic Management Journal, 9*(4), 319–332.

Laumann, E. O., Galaskiewicz, J., & Marsden, P. V. (1978). Community structure as interorganizational linkages. *Annual Review of Sociology, 4*(1), 455–484.

Lawrence, T. B. (1999). Institutional strategy. *Journal of Management, 25*(2), 161–187.

Maguire, S., Hardy, C., & Lawrence, T. (2004). Institutional entrepreneurship in emerging fields: HIV/AIDS treatment advocacy in Canada. *Academy of Management Journal, 47*(5), 657–679.

Meyer, R. E., & Höllerer, M. A. (2014). Does institutional theory need redirecting?. *Journal of Management Studies, 51*(7), 1221–1233.

Meyer, J. W., & Rowan, B. (1977). Institutionalized organizations: Formal structure as myth and ceremony. *American journal of sociology, 83*(2), 340–363.

Meyer, J. W., & Scott, W. R. (1983). *Organizational environments: Ritual and rationality.* Thousand Oaks, CA: Sage.

Miles, R. E., & Snow, C. C. (1986). Organizations: New concepts for new forms. *California Management Review, 28*(3), 62–73.

Mitchell, W. (1969). *Social networks in urban situations.* Manchester, England: Manchester University Press.

Morgan, K. (2004). The exaggerated death of geography: Learning, proximity and territorial innovation systems. *Journal of economic geography, 4*(1), 3–21.

Nonaka, I., & Takeuchi, H. (1995). *The knowledge creating company.* Oxford, England: Oxford University Press.

North, D. C. (1990). *Institutions, institutional change and economic performance.* Cambridge, England: Cambridge University Press.

North, D. C. (1993). Institutions and credible commitment. *Journal of Institutional and Theoretical Economics (JITE)/Zeitschrift für die gesamte Staatswissenschaft, 149*(1), 11–23.

Oliver, C. (1990). Determinants of interorganizational relationships: Integration and future directions. *Academy of Management Review, 15*(2), 241–265.

Oliver, C. (1991). Strategic responses to institutional processes. *The Academy of Management Review, 16*(1), 145–179.

Orsenigo, L., Pammolli, F., & Riccaboni, M. (2001). Technological change and network dynamics Lessons from the pharmaceutical industry. *Research Policy, 30*(3), 485–508

Orsenigo, L., Pammolli, F., Riccaboni, M., Bonaccoris, A., & Turchetti, G. (1998). The evolution of knowledge and the dynamics of an industry network. *Journal of Management & Governance, 1*(2), 147–175.

Ozman, M. (2009). Inter-firm networks and innovation: a survey of literature. *Economics of Innovation and New Technology, 18*(1), 39–67.

Patel, P., & Pavitt, K. (1994). The continuing, widespread (and neglected) importance of improvements in mechanical technologies. *Research Policy, 23*(5), 533–545.

Peng, M. W., Sun, S. L., Pinkham, B., & Chen, H. (2009). The institution-based view as a third leg for a strategy tripod. *Academy of Management Perspectives, 23*(3), 63–81.

Peng, M. W., Wang, D. Y., & Jiang, Y. (2008). An institution-based view of international business strategy: A focus on emerging economies. *Journal of International Business Studies, 39*(5), 920–936.

Pennings, J. M. (1981). Strategically interdependent organizations. *Handbook of Organizational Design, 1,* 433–455.

Pennings, J. M., & Harianto, F. (1992a). The diffusion of technological innovation in the commercial banking industry. *Strategic Management Journal, 13*(1), 29–46.

Pennings, J. M., & Harianto, F. (1992b). Technological networking and innovation implementation. *Organization Science, 3*(3), 356–382.

Perks, H., & Jeffery, R. (2006). Global network configuration for innovation: A study of international fibre innovation. *R&D Management, 36*(1), 67–83.

Pisano, G. P. (1991). The governance of innovation: vertical integration and collaborative arrangements in the biotechnology industry. *Research Policy, 20*(3), 237–249.

Podolny, J., & Page, K. (1998). Network forms of organization. *Annual Review of Sociology, 24,* 57–76.

Powell, W., Koput, K., Smith-Doerr, L., & Owen-Smith, J. (1999). Network position and firm performance: Organizational returns to collaboration in the biotechnology industry. *Research in the Sociology of Organizations: A Research Annual, 16,* 129–159.

Powell W. W., Koput, K. W., & Smith-Doerr, L. (1996). Interorganizational collaboration and the locus of innovation: Networks of learning in biotechnology. *Administrative Science Quarterly, 41*(1), 116–145.

Powell, W. W., White, D. R., Koput, K. W., & Owen-Smith, J. (2005). Network dynamics and field evolution: The growth of interorganizational collaboration in the life sciences1. *American Journal of Sociology, 110*(4), 1132–1205.

Schumpeter, J. A. (1934). *The theory of economic development: An inquiry into profits, capital, credit, interest, and the business cycle.* New Brunswick, NJ: Transaction Books.

Scott, W. R. (1987). *Organizations: Rational, natural, and open systems.* Englewood Cliffs, NJ: Prentice-Hall.

Scott, W. R. (1995). *Institutions and organizations: Foundations for organizational science.* Thousand Oaks, CA: Sage.

Shan, W., Walker, G., & Kogut, B. (1994). Interfirm cooperation and startup innovation in the biotechnology industry. *Strategic Management Journal, 15*(5), 387–394.

Sweeney, G. P. (1995). *National innovation policy or a regional innovation culture.* Working Papers in European Industrial Policy, no. 1. European Network on Industrial Policy.

Taggart, J. (2007). Determinants of the foreign R&D locational decision in the pharmaceutical industry. *R&D Management, 21*(3), 229–240.

Teece, D. J. (1992). Strategies for capturing the financial benefits from technological innovation. In N. Rosenberg, R. Landau, & D. C. Mowery (Eds.), *Technology and the wealth of nations* (pp. 509–533). Stanford, CA: Stanford University Press.

Uzzi, B. (1997). Social structure and competition in interfirm networks: The paradox of embeddedness. *Administrative Science Quarterly, 42*(1), 35–67.

Walker, G., Kogut, B., & Shan, W. (1997). Social capital, structural holes and the formation of an industry network. *Organization Science, 8*(2), 109–125.

Zukin, S., DiMaggio, P. (1994). *Structures of capital: The social organization of the economy.* Cambridge, England: Cambridge University Press.

CHAPTER 2

STAKEHOLDER CENTRALITY POTENTIAL AND RELATIONAL IDENTIFICATION IN TECHNOLOGICAL INNOVATION NETWORKS

Israël Fortin
Erasmus University

ABSTRACT

Interest-based motivations predict stakeholders sharing the same interests will likely act similarly within a social network. Contrariwise, identity-based motivations forecast similar stakeholders will feel threatened by each other and rather behave differently in order to remain distinct from one another in a social network. Although identity-motivated behaviors can appear irrational from an interest-based perspective, both perspectives are complementary in explaining the complexity encountered when analyzing interactions occurring in technological innovation networks, where knowledge recombination is believed to trigger innovation despite complex supply chain relationships. Identities are commonly approached as a simple antecedent to innovation in the literature,

Technological Innovation Networks, pages 19–45

nurturing or deterring initiatives. Through an original and dynamic typology of centrality potential based on relational identifications among stakeholders of technological innovation network, we rather aim to describe how identities and identifications may predict future stakeholder behaviors.

Just as with identities, innovations can hardly be forced out of stakeholders. Instead, managers gather available resources around potential innovators, provide incentives, and wait with impatience. Since organizations do not always possess all relevant resources for innovation, participation in technological innovation networks overcomes the lack of resources through strategic alliances on research and development projects, which also distribute financial risks among partners (Brass, Galaskiewicz, Greve, & Tsai, 2004). Consequently, relationships with central stakeholders in technological innovation networks might be the most valuable resources an organization develops regardless of its size and stage of evolution. However, genuine relationships are based on genuine identity overlaps defining the strength of identification.

Identification with central stakeholders orchestrating formal or informal technological innovation networks might catalyze successful innovation projects or be its driving force. This chapter first reviews the literature on stakeholder typologies in order to bring an important distinction between "static" and "dynamic" typologies. Second, studies about relationships between stakeholders are presented. Two axes defining eight ways to study stakeholder relationships report the evolution of research from dyadic relationships centered on a focal organization toward stakeholder networks focusing on several relationships. Finally, we review the literature on identity in stakeholder networks in order to suggest an original typology of centrality potential based on relational identification with central stakeholders, initially created as a tool to analyze interactions in a research consortium in aerospace. This typology presents five dynamic stakeholder attitudes and suits to several levels of analysis, consistent with Freeman's (1984) seminal definition of a stakeholder being any group or individual who can affect, or is affected by, the achievement of an organization's purpose.

Relational identification combines role-based identification and person-based identification. This psychology concept has been applied in the organization sciences at the individual level (Sluss & Ashforth, 2007) and at the group level (Hogg, Knippenberg, & Rast, 2012) with suggested convergence with other identities in which they are "nested" or "embedded" such as organizational identity (Sluss & Ashforth, 2008; Sluss, Ployhart, Cobb, & Ashforth, 2012). Rather than simply identifying with a specific role or a specific stakeholder, stakeholders identify with roles they play in relationships with specific stakeholders (individuals or organizations). This simple definition opens a variety of identification possibilities including disidentification and ambivalent identification, meaning exclusively role-based or solely stakeholder-based identification. The perception of one's role, the number of stakeholders one experiences a role with, and the number of different roles a stakeholder fulfills towards another also affect the relationship, the identification, and the generalization of the relationship to similar stakeholders. A dynamic typology of stakeholders based on relational identification, disidentification, and ambivalent identification is thus offered and linked to previous existing typologies.

STAKEHOLDER TYPOLOGIES: THE STATIC OR DYNAMIC PERCEPTION OF STAKEHOLDERS

Since Freeman (1984), several literature reviews about stakeholders have been published (Heath & Norman, 2004; Laplume, Sonpar, & Litz, 2008; Mainardes, Alves, & Raposo, 2011). These works clearly show the preponderance of managerial perceptions in the identification of stakeholders and promote categories following environment changes. Yet theories first considered a limited range of stakeholders, largely predetermined, common to all organizations, and mostly corporations. More recent works conceive an extended number of stakeholders, increasingly abstract, more dynamic, and admit the organic emergence of new stakeholder categories. Some scholars disapprove overly inclusive definitions of generic stakeholders and would prefer a consensus on tighter definitions of specific stakeholders (Fassin, 2009), while others question the necessity for managers to pay attention to all possible stakeholders, actual or anticipated future ones (Mainardes, Alves, & Raposo, 2011). This critique has been in part addressed by the theory of salience (Mitchell, Agle, & Wood, 1997), considering stakeholders from the organization's point of view (through officials), according to their perceived power, urgency, and legitimacy. Moreover, stakeholder salience would follow the development cycle of the organization (Jawahar & McLaughlin, 2001). The present review tracks two distinct perceptions of stakeholders: first, static typologies approaching the perception of stakeholders as stable in time, and then dynamic typologies where the perception of stakeholders varies with the unfolding of time and events.

Static Perceptions of Stakeholders

Table 2.1 offers an overview of static typologies proposed so far and does not pretend to include all of them. Static typologies present stakeholder categories as stable in time and often offer binary categories among which one category is considered more important or more stable than the other.

TABLE 2.1 Examples of Typologies Adopting a Static Perception of Stakeholders

Authors	Stakeholder Categories
Freeman, R. E. (1984). *Strategic management: A stakeholder approach.* Boston, MA: Pitman.	Primary stakeholders Secondary stakeholders
Clarkson, M. B. E. (1995). A stakeholder framework for analysing and evaluating corporate social performance. *Academy of Management Review, 20*(1), 92–117.	Primary stakeholders Public stakeholders
Henriques, I., & Sadorsky, P. (1999). the relationship between environmental commitment and managerial perceptions of stakeholder importance. *Academy of Management Journal, 42*(1), 87–99.	Regulatory stakeholders Organizational stakeholders Community stakeholders The media
Post, J. E., Preston, L. E., & Sachs, S. (2002). Managing the extended enterprise: The new stakeholder view. *California Management Review, 45*(1), 6–28.	Resource base Industry structure Social political arena
Hart, S. L., & Sharma, S. (2004). Engaging fringe stakeholders for competitive imagination. *Academy of Management Executive, 18*(1), 7–18.	Core stakeholders Fringe stakeholders
Fanelli, A., & Misangyi, V. F. (2006). Bringing out charisma: CEO charisma and external stakeholders. *Academy of Management Review, 31*(4), 1049–1061.	Internal stakeholders External stakeholders
Kacperczyk, A. (2009). With greater power comes greater responsibility? Takeover protection and corporate attention to stakeholders. *Strategic Management Journal, 30,* 261–285.	Shareholders Non-shareholding stakeholders
Choi, J., & Wang, H. (2009). Stakeholder relations and the persistence of corporate financial performance. *Strategic Management Journal, 30,* 895–907.	Financial stakeholders Nonfinancial stakeholders
Lawrence, A. T. (2010). Managing disputes with nonmarket stakeholders: Wage a fight, withdraw, wait, or work it out. *California Management Review, 53*(1), 90–113.	Market stakeholders Nonmarket stakeholders

Typologies of stakeholders perceived as static are often illustrated using concentric circles where the closest to the core a stakeholder category appears, the more important, relevant, or critical it is assumed to be for the organization. The most prevalent static typologies of stakeholders oppose "primary" or "internal" stakeholders to "secondary" or "external" stakeholders. This way, some stakeholders are considered permanently "critical," such as owners (stockholders or shareholders), managers, or clients, and the centrality of these stakeholders is believed to remain constant. However, new realities brought by globalization and communication technology developments eventually renewed these typologies.

As previously mentioned, Edward Freeman (1984) proposed an influential definition as well as one of the most largely mobilized stakeholder typologies so far. According to Freeman (1984), "a stakeholder in an organization is (by definition) any group or individual who can affect or is affected by the achievement of the organization's objectives" (p. 46). This initial definition holds the potential to include numerous stakeholders, but Freeman defines two large categories usually relevant for all organizations. Primary stakeholders are common to all organizations and usually encompass investors, employees, consumers, and suppliers. Secondary stakeholders usually include governments, unions, communities, NGOs, and so on. In principle, stakeholder theory moves away from a mainstream hierarchical conception of firms by putting stakeholders on an equal ground, but the sustained interest of researchers in relationships with stakeholders located on top of the traditional hierarchy, in particular stockholders, perpetrates the former hierarchical vision. Other terms ultimately refer to primary and secondary stakeholder categories. For instance, definitions of "market" and "nonmarket" stakeholders or "financial" and "nonfinancial" stakeholders are fairly equivalent to Freeman's primary and secondary stakeholders, respectively (Lawrence, 2010, p. 92).

Some authors defend the relevance of a new vision of stakeholders (Post, Preston, & Sachs, 2002), justified by the changes brought by globalization (Jensen & Sandström, 2011) and the new possibilities of information and communication technologies, essentially the Internet. Jensen and Sandström (2011) believe globalization challenges stakeholder theory in two ways: "The first challenge is to acknowledge new power relations (sub-political movements, news forms of bureaucracy and hierarchy), and the second is to acknowledge new dimensions of responsibility (a political responsibility)" (p. 473). Facing these changes, some alternative definitions were suggested. For instance, Post et al. (2002) define stakeholders as "individuals and constituencies that contribute, either voluntarily or involuntarily, to [an organization's] wealth-creating capacity and activities, and who are therefore its potential beneficiaries and/or risk bearers" (p. 8).

Post et al. (2002) renamed Freeman's initial categories and added a middle group. For them, stakeholders either belong to the resource base, the industry structure, or the social political arena. This typology contrasts with Hart and Sharma's (2004) typology, which counts only two categories, core or fringe stakeholders. These authors rather consider former "secondary" and "primary" stakeholders as "core" to the organization and qualify all other stakeholders as "fringe," while stressing their importance in the long-term survival of the organization. We can, however, debate the real contribution of Post et al. and Hart and Sharma when comparing their typologies with previous ones. It is noteworthy that Clarkson's (1995) typology presents an economy of terms in considering stakeholders according to their access to information. In fact, it is reasonable to believe stakeholders considered external to the organization access similar public information, allowing the quick and unprecedented mobilization of any interest group, including some unsuspected ones. In many cases, a stakeholder corresponds to a general public (consumers, voters, society, the community, the collectivity, citizens, etc.), more or less limited (municipal, provincial, national, international), with or without the mediation of journalists and their media (press, television, Internet, etc.). Cases of organizations having the public as a stakeholder are often about multinationals with international consumers, distributors, suppliers, employees, and above of all stockholders, which explains the concentrations of studies around these organizations (Laplume et al., 2008).

In summary, since Freeman (1984), several typologies have been offered in which the perceived importance of various stakeholder categories is presumed stable in time. In particular, Post et al. (2002) suggest apprehending stakeholders according to an extended enterprise perspective, which includes the resource base, the industrial structure, and the sociopolitical arena. Hart and Sharma (2004) extend even further the scope of stakeholders, including fringe stakeholders, ignored so far by organizations as well as by scholars (week, poor, divergent, adversarial, isolated groups, etc.). Nevertheless, these stakeholders appear more or less "salient" to managers (Jones, Felps, & Bigley, 2007; Mitchell et al., 1997) and mold organizational strategies in significant and often unexpected ways, as in many well-documented cases (Nike, Shell, Monsanto, Nestlé, BP, etc.).

Dynamic Perceptions of Stakeholders

Mitchell et al. (1997) are the first to define the prime importance of managerial perceptions in stakeholder identification, giving birth to an important stream of literature. Mitchell et al. first classify stakeholders according to their salience to managers, in terms of power, legitimacy, and urgency. These three dimensions further define eight types of stakeholder exposing none, one, two, or three of those attributes: dangerous, definitive, dependent, dormant, discretionary, demanding, and non-stakeholders. The centrality of a stakeholder varies with the perception of decision makers. Therefore, if these perceptions change, stakeholders may change categories. Among others, Mitchell et al.'s (1997) typology has been revisited by Jones et al. (2007), Eesley and Lenox (2006), and Jawahar and McLaughlin (2001). Jones et al. add an organizational culture dimension to stakeholder salience, which follows an egotistical, instrumental, or moralist orientation. Eesley and Lenox (2006) define salience rather in action terms and propose that stakeholder power, legitimacy, and urgency emerge from the triad defined by the stakeholder's nature, its request, and the organization. Lastly, Jawahar and McLaughlin (2001) suggest that stakeholder salience naturally changes with the organization evolution according to a predetermined and predictable cycle. Again, environment changes soon forced the evolution of the newly established bases of the salience theory.

Global technology changes brought Harrison and Freeman (1999) to request more theoretical conceptions in order to "create rich and rigorous cases that could lead us to see the overall stakeholder relationship as a multifaceted, multiobjective, complex phenomenon" (p. 484). Table 2.2 presents such dynamic typologies.

TABLE 2.2 Examples of Typologies Adopting a Dynamic Perception of Stakeholders

Authors	Stakeholder Categories
Savage, G. T., Nix, T. W., Whitehead, C. J., & Blair, J. D. (1991). Strategies for assessing and managing organizational stakeholders. *Academy of Management Executive, 5*(2), 61–75.	Supportive stakeholders Marginal stakeholders Nonsupportive stakeholders Mixed blessing stakeholders
Mitchell, R. K., Agle, B. R., & Wood, D. J. (1997). Toward a theory of stakeholder salience: Defining the principle of who and what really counts. *Academy of Management Review, 22*, 853–886.	*Latent stakeholders* (dormant, discretionary and demanding stakeholders) *Expectant stakeholders* (dominant, dependent, and dangerous stakeholders) *Definitive stakeholders* *Nonstakeholders*
Rowley, T. J. (1997). Moving beyond dyadic ties: A network theory of stakeholder influences. *Academy of Management Review, 22*(4), 887–910.	Compromiser Commander Subordinate Solitarian
Jawahar, I. M., & McLaughlin, G. L. (2001). Toward a descriptive stakeholder theory: An organizational life cycle approach. *Academy of Management Review, 26*(3), 397–414.	Critical stakeholders Other stakeholders
Friedman, A., & Miles, S. (2002). Developing stakeholder theory. *Journal of Management Studies, 39*(1), 1–21.	Explicit recognized Implicit unrecognized Implicit recognized No contract
Luk, C.-L., Yau, O. H. M., Tse, A. C. B., Sin, L. Y. M., & Chow, R. P. M. (2005). Stakeholder orientation and business performance: The case of service companies in China. *Journal of International Marketing, 13*(1), 89–110.	Synergistic stakeholders Hindering stakeholders
Fassin, Y. (2009). The stakeholder model refined. *Journal of Business Ethics, 84*, 113–135.	Stakeholders Stakewatchers Generic stakekeepers Specific stakekeepers Nonstakeholders

An interesting contribution of typologies adopting a dynamic perception of stakeholders—in comparison to typologies adopting a static perception of stakeholders—lies in leaving behind internal and external stakeholder distinctions for more blurry organizational boundaries. Some stakeholder categories, for instance workers' unions, have always been a challenge to classify as either internal or external according to traditional typologies. Some organization types such as cooperatives or public societies present configurations that further blur organizational boundaries in terms of internal or external stakeholders. On the other hand, these typologies confirm the central role played by stakeholder groups perceiving and judging other stakeholders on behalf of the organization, namely decision makers, upper management, directors, and so on. Dynamic stakeholder typologies discard concentric circle representations and often present tables less prone to schematization. We will visit the typologies of Savage, Nix, Whitehead, and Blair, (1991) and Rowley (1997) in more detail subsequently to help us build and discuss our own dynamic typology.

To sum up, most organizations consider their stakeholders as being either primary and central or secondary and peripheral, in a predetermined fashion in line with their corporate strategy, and rationally consider them more or less critical or important. Although reciprocal relationships were initially considered between an organization and its stakeholders, researchers focused first on relationships from the organization's point of view, with a limited number of stakeholders, often a single one (Rowley, 1997). While stakeholder typologies proliferate and are refined, conceptions of the relationships between an organization and its stakeholders also develop in parallel. In particular, as early as 1997, Rowley suggested extending stakeholder studies beyond dyadic relationships to consider multiple interrelations between organizations and their stakeholders.

STAKEHOLDER RELATIONSHIPS: FROM DYADS TO NETWORKS

Studying relationships between stakeholders implicates a set of choices delimiting the scope of the relationships under investigation and underlying important methodological consequences. Whether researchers consider the perspective of a single stakeholder in a dyad or concentrate on the interrelations within an entire network, they define a specific angle along with appropriate methods. The attention on a single dyad orients projects toward qualitative cases, whereas the study of an entire network agrees with quantitative research designs. Following a review of works about relationships in organization sciences, we circumscribed three ways to study relationships between stakeholders: simple dyadic, multiple dyadic, or network

TABLE 2.3 Relationships Between Stakeholders (SH)

Perspective	Dyadic Relationships: Simple	Dyadic Relationships: Multiple	Network of Relationships
Perception by a focal SH	Simple dyadic relationship perceived by a focal SH	Multiple dyadic relationships perceived by a focal SH	Network perceived by a focal SH (egonetwork)
Perceptions of a focal SH by others		Multiple dyadic relationships of a focal SH perceived by others	Perceptions of a focal SH by the network
Bidirectional	Simple dyadic interrelations	Multiple dyadic interrelations	Network of interrelations

relationships. Within each of these research strategies, researchers adopt three perspectives: from the perspective of a stakeholder, from the perspective of all stakeholders, or from the perspective of several stakeholders towards one specific stakeholder (Table 2.3). First of all, studies of a simple dyadic relationship focus on the relationship defined by two stakeholders. Second, studies of multiple dyadic relationships are interested by several relationships defined by several dyads sharing a common focal stakeholder. Finally, network studies are concerned by triads of relationships among several stakeholders. We now turn to a brief overview of the literature on simple and multiple dyadic relationships to further explore in detail relationships in networks.

Dyadic Relationships

Studies on dyadic relationships—meaning relationships between two stakeholders, whether individuals or collectives—surge by hundreds every year in organization sciences. One can only recall articles about the relationship between an organization and its CEO, director board, upper management, stockholders, shareholders, employees, clients, suppliers, local communities, governments, NGOs, media, and so on. Yet scholars often focus on the perspective of one individual or collective stakeholder in these dyads. This concentration of research on simple dyadic relationships remains a frequent critique addressed to the stakeholder literature (Frooman, 1999; Mainardes et al., 2011; Mitchell et al., 1997; Neville & Menguc, 2006; Rowley, 1997).

Other scholars study the points of view of both stakeholders forming the dyad. Therefore, their works concern simple dyadic interrelations (Figure 2.1). Among others, the emphasis on simple dyadic relationships is typical of interorganizational relationships.

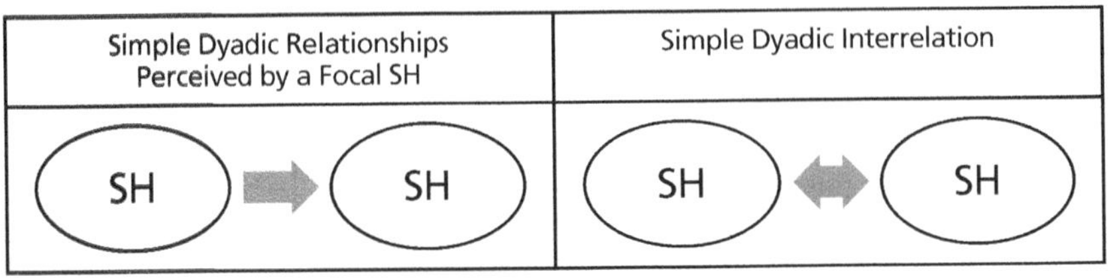

Figure 2.1 Simple dyadic relationships between stakeholders (SH).

Some authors are interested by relationships involving more than two actors but maintain their attention on a single focal stakeholder. Just as the investigation of simple dyadic relationships, the study of multiple dyadic relationships considers each dyad in isolation. Yet this research stream faces two scenarios: either to focus on a limited number of collective actors, or to restrict the study to few key individuals. Figure 2.2 schematizes the three possible categories of multiple dyadic relationships. The study of multiple dyadic relationships is typical of the literature stream on multiple stakeholders. For instance, and as mentioned previously, a major domain of the stakeholder literature consists in identifying and prioritizing the stakeholders of a focal organization according to their salience (Mitchell et al., 1997), following different dimensions.

Studies of multiple dyadic relationships compare interrelations between a focal stakeholder and other stakeholders, different perceptions of a focal stakeholder by other stakeholders, or how a focal stakeholder perceives its relationships with others. However, stakeholder dyads do not evolve in isolation and these studies overlook possible triads of interrelations. According to Rowley (1997) "since stakeholder relationships do not occur in a vacuum of dyadic ties, but rather in a network of influences, a firm's stakeholders are likely to have direct relationships with one another" (p. 890). Today,

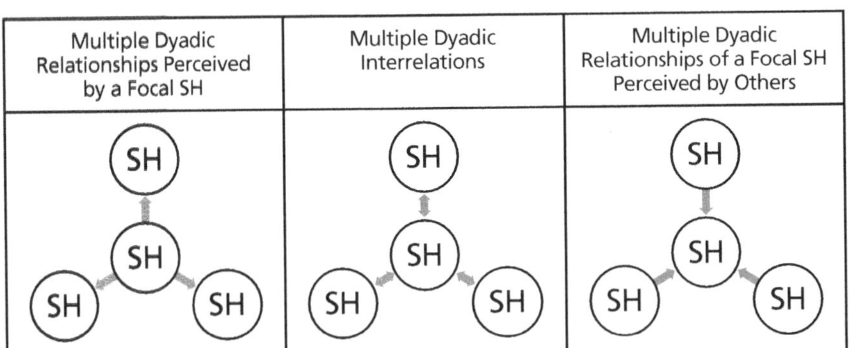

Figure 2.2 Multiple dyadic relationships between stakeholders (SH).

not only do authors report interrelations between a focal stakeholder and other stakeholders, but they also abandon the attention granted to a single focal actor in these interrelations and conceive triads of interrelations as well as networks of stakeholders.

Networks of Relationships

Rowley (1997) seems the instigator of a research domain focusing on stakeholder networks. Although stakeholder and network literatures share a relational perspective, this expression remains problematic since both approaches appear difficult to reconcile: "Network models begin where stakeholder research stops—the dyadic relationship—and examine systems of dyadic interactions, capturing the influence of multiple and interdependent relationships on organizations' behaviors" (p. 894). In addition to expanding the focus on relationships between stakeholders, the network perspective clearly delimits either organizational or personal networks and avoids blending individual and collective stakeholders as it is possible to do (albeit rarely done) with a stakeholder perspective. Yet, "levels-of-analysis issues are especially relevant for network researchers, whose research questions usually lie at the intersection between different levels" (Capaldo, 2007, p. 588). Moreover, macro and micro levels may be theoretically and methodologically very similar (Borgatti & Foster, 2003), which is why theories and constructs initially developed for interpersonal relationships are frequently applied to the analysis of interorganizational ties (Capaldo, 2007).

Contrary to studies of dyadic relationships, studies of social networks face great challenges in considering each possible dyad, unless they only focus on three or four actors or only on individuals and avoid collectives. It is, however, possible to investigate a network as a more or less homogenous whole by considering only simple and specific dimensions or relationships, which leans toward quantitative approaches. An important distinction between multiple dyadic relationships and a network perspective lies in the appreciation of the relationships among all actors simultaneously rather than isolated dyads centered on a focal actor. Figure 2.3 illustrates such network configurations.

Scholars investigating a stakeholder perception by a network are typical of studies on organizational reputation, whereas studies about the perception of a network by a focal actor are traceable in the literature about multi-stakeholder partnerships as well as in the literature on ego-networks. Finally, networks of interrelations are typical of studies on networks.

Following these brief introductions of stakeholder typologies and stakeholder relationships, we now turn to works on identity in stakeholder networks in order to define a dynamic stakeholder typology acknowledging

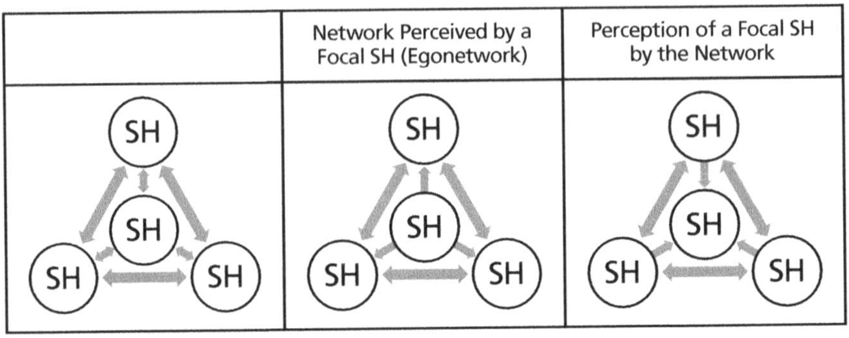

Figure 2.3 Relationships in networks of stakeholders (SH).

identifications of stakeholders among each other within a technological innovation network which would predict the "centrality potential" of some stakeholders' categories evolving in these networks.

IDENTITY IN STAKEHOLDER NETWORKS

Although most stakeholder definitions refer to collective as well as individual actors, scholars most frequently use the term in reference to collectives. Therefore, research on identity in stakeholder relations relate to collective identities, mostly organizational identities. For this reason, we first review the literature on organizational identity and image, and later move to the more specific concept of relational identity, which will be used in the conception of a dynamic stakeholder typology.

Identities, Images and Networks

From the organization's point of view, organizational images constructed by internal and external stakeholders are preoccupying since they contribute to the organizational identity definition. The definition of organizational identity designating what is central, distinctive and enduring of an organization (Albert & Whetten, 1985) is without a doubt the most widely recognized one (Brown, 2001). Yet the enduring aspect of organizational identity has been reconsidered "adaptatively unstable" (Gioia, Schultz, & Corley, 2000) and "sticky" (Scott & Lane, 2000). These nuances add a paradoxical dimension to organizational identity that appears central, distinctive, and enduring, as well as adaptable to changes in the environment. The concept of organizational image further complexifies this paradox.

> An organization's identity describes what its members believe to be its character; an organization's image describes attributes members believe people outside the organization use to distinguish it. Organizational image is different from reputation: reputation describes the actual attributes outsiders ascribe to an organization, but image describes insiders' assessments of what outsiders think. Both organizational image and identity are constructs held in organization members' minds. (Dutton & Dukerich, 1991, p. 547)

Organizational identity evolves with time (Gioia & Thomas, 1996) and exists among members of an organization at the individual as well as at the collective levels (Dutton, Dukerich, & Harquail, 1994). At the individual level, organizational identity is an interpretation unique to each individual that "may not match a collective organizational identity that represents the members' shared beliefs" (Dutton et al., 1994, p. 240). This is why at the individual level, Dutton et al. (1994) refer instead to "perceived organizational identity" and use the term "organizational image" for "what the member believes is central, distinctive, and enduring about the organization" (p. 239).

As opposed to organizational identity, the definition of organizational image does not seem to benefit from a consensus. Dutton and Dukerich (1991) used it to designate what organizational members believe to be the external perception of the organization. Later, Dutton et al. (1994) renamed this conception "construed external image." For Alvesson (1990), organizational images are created by the organization for outside communication. Gioia and Thomas (1996) also distinguish a "future desired image"—different from the actual image but similar to reputation (internal or external). Organizational image, integrating elements internal and external to the organization, is referred to as "projected image" by Gioia et al. (2000), in between "future desired image" and corporate identity: "Projected images typically are associated with specific contexts, events, issues, and audiences" (p. 66). Finally, Gioia et al. (2000) defined organizational image as "a wide-raging concept connoting perceptions that are both internal and external to the organization, as well as perceptions that are both projected and received" (p. 66).

Hatch and Schultz (1997, 2002) are noteworthy among authors conceptualizing organizational image and identity. In a first 1997 model, authors clearly locate organizational identity as a co-construction within organizational culture, while image is externally co-constructed. Only two groups appear involved in both processes: leaders and top managers. Leaders and top managers are implicated in organizational identity co-construction in concert with employees and in organizational image co-construction with external groups. However, a second model from Hatch and Schultz (2002) shows organizational identity as negotiated between organizational internal culture and external image without the intervention of leaders or top managers. It remains that organizational image often appears external and organizational identity internal to the organization.

Some scholars have looked at the construction and formation of identities involved in the interaction of members from different organizations (Zhang & Huxham, 2009). Recently, some authors (Brickson, 2005; Pratt & Foreman, 2000; Randel, Jaussi, & Standifird, 2009; Scott & Lane, 2000) established a relation between organizational identity and stakeholders and suggested "re-reading models of the relational nature of organizational identity exploring the interrelations with the stakeholder approaches" (Illia & Lurati, 2006, p. 300). Scott and Lane (2000) seem to be the first to propose "a manager–stakeholder view that specifically addresses organizational identity as emerging from the complex interactions among managers, organizational members, and other stakeholders" (p. 44). The construction of organizational identity occurs among top management communicating desired organizational images intended for stakeholders. In return, stakeholders formulate "appraisals," reflecting desired organizational images directed towards top managers. This process gives birth to an organizational identity located in the communications between top management and organizational stakeholders.

According to Scott and Lane, organizations mobilize three tactics to amplify the identification of "member" and "non-member" organizational stakeholders: "1) organizational communications, 2) the enhancement of the visibility of stakeholders' organizational affiliations, and 3) the embedding of stakeholders within the organizational community" (2000, p. 58). Surprisingly, Scott and Lane also warn against interaction between an organization and its stakeholders:

> In trying to manage identification, organizations may unwittingly boost stakeholder power vis-à-vis managers by encouraging interaction within or among stakeholder groups.... By increasing interaction among different stakeholder groups, an organization may decrease its centrality in the overall stakeholder network. (2000, p. 58)

In a rare empirical study on 1126 participants from 88 organizations (legal services and nonalcoholic beverage industry) using quantitative and qualitative methods, Brickson (2005) concludes that the orientation of an organizational identity (individualist, relational or collectivist) is strongly dictated by the relationships between an organization and its stakeholders: "[S]pecific organizational variables bearing on organizations' stakeholder relations appear to be stronger predictors of identity orientation than general organization-level or individual-level variables." For Randel et al. (2009), organizational identity filters negative evaluations from stakeholders perceived as identity threats.

Relational and interpersonal identity dimensions are greatly privileged by the rare studies investigating the links between organizational identity and stakeholders. A strong empirical trend also converges several personal and collective identity concepts at the interpersonal level of analysis (Alvesson, Ashcraft, & Thomas, 2008; Ashforth, Harrison, & Corley, 2008; Prentice, 2001). This is the case for "relational" identity, encompassing personal and role identity. Within their different relationships maintained with specific stakeholders, organizational representatives fulfill distinct roles. For instance, a CEO will act differently with a new union representative, a recently layoff employee, a long-time supplier, a consumers association, a deputy minister, or a potential acquirer. Relational identity accounts for these differences at the interactional level. Moreover, Sluss and Ashforth (2008) suggest both role and organizational identities mutually reinforce each other: "It seems logical that role relationships and organizations, as stimuli for identification, may 'resemble' each other because they are structurally nested entities" (p. 811).

Relational Identity

Several organizational sciences scholars follow social psychologists (Prentice, 2001; Tice & Baumeister, 2001) in their efforts to converge personal and collective identities at the interpersonal level (Ashforth & Sluss, 2006; Chen & Andersen, 2008; Sluss & Ashforth, 2007). Relational identity is born from one of these convergences, grouping personal and role identities.

Alvesson et al. (2008) note: "[A]s theories of the inescapable personal-social relation are currently applied to empirical studies of self-identity, we notice a tendency toward highly localized notions of relational context and interactive production" (p. 11). The inevitable relation between the social and the personal inside identity configurations "does not entail simply stepping into pre-packaged selves, but always involves negotiating intersection with the people and systems around us" (Alvesson et al., 2008, p. 10).

In psychology, relational self refers to features affected by close relationships with a significant other, presenting little interest for organizational sciences. But the prime focus on relational identity presents individual and role identities as elements of relational identity that extend beyond close relationships to include individuals generalized to significant groups. This last conception found in organizational research is mobilized in this chapter. Brewer and Gardner (1996) seem to be the first to conceive relational identity this way. Among others, their model has been applied by Sluss and Ashforth (2007) and Brickson (2000).

In particular, Sluss and Ashforth slightly modify Brewer and Gardner's terms and bring an important nuance: "It is not accurate to use the term levels of analysis, because the interpersonal and collective levels refer here to an individual's conception of his or her relational and role-based identities rather than to identities apart from the individual" (2007, p. 13). In fact, although these authors refer to three distinct levels of analysis, the three described identities relate to three different identity conceptions referring to three different perceptions maintained by individuals.

> We define a relational identity as the nature of one's role-relationship, such as manager-subordinate and coworker-coworker. It is how role occupants enact their respective roles vis-à-vis each other. As noted above, a role is fundamentally relational and is largely understood with reference to the network of interdependent roles. (Sluss & Ashforth, 2007, p. 11)

Relational identity is composed of four components: personal and role identities of oneself, and personal and role identities of others. "A role-based identity is the goals, values, beliefs, norms, interaction styles, and time horizons typically associated with the role—independent of who may be enacting the role" (Sluss & Ashforth, 2007, p. 11). This interpersonal perspective integrates individual and collective identities through person- and role-based identities, respectively. "A relational identity, however, focuses on that portion of the role-based identity that is more or less directly relevant to one's role-relationship" (Sluss & Ashforth, 2007, p. 11). Furthermore, individuals involved in the relationship dispose of some latitude in enacting their role identity. "The person-based identity shapes the enactment of the role-based identity (and vice versa)" (Sluss & Ashforth, 2007, p. 11).

Sluss and Ashforth's (2007) model of relational identity mobilizes evaluative and appreciative variables and also permits emotional ambivalence. Figure 2.4 draws three possible scenarios resulting from the crossing of two axes: positive or negative identification (disidentification) to role and personal identities (the two identities concerned by relational identity). "Relational identification occurs if the valence of the partner's role- and person-based identities is positive, relational disidentification occurs if the valence

		Person-based identity	
		Negative	Positive
Role-based identity	Positive	Ambivalent relational identification	Relational identification
	Negative	Relational identification	Ambivalent relational identification

Figure 2.4 A typology of relational identification (*Source:* Sluss & Ashforth, 2007).

is negative, and ambivalent relational identification occurs if the valence is mixed" (Sluss & Ashforth, 2007, p. 25).

Ashforth and Sluss (2006) had previously described three unhealthy scenarios also possible in relational identification, namely overidentification, underidentification, and relational identity transfer. The simple typology of Sluss and Ashforth may, however, get quickly complex when considering relationships occurring among various actors in a stakeholder network and when considering several levels of analysis. Assuming that identity overlaps combine with interest overlaps in explaining actors' behaviors in a network (Rowley & Moldoveanu, 2003), we are now moving to a typology of relationally identified stakeholders in a network, considering interactions among only three types of stakeholders according to their centrality. We will then show how the typology may apply to several levels of analysis.

RELATIONAL IDENTITY IN TECHNOLOGICAL INNOVATION NETWORKS: A TYPOLOGY

Despite the impressive number of stakeholder typologies available, none of them accounts for the complexity of the dynamic stakeholder relationships encountered in technological innovation networks typically underlying a supply chain and aiming at innovation. The dynamic typology suggested in this chapter might help managers and analysts to situate organizations in relation to one another in an ever-evolving technological innovation network, aid them in evaluating the potential of a position in the network, and develop strategies to reach better positions while remaining faithful to their identity. For instance, a research consortium in aerospace, underlying supply chain interactions between stakeholders embedded in a wider technological innovation network including industrial, academic, and government partners, would appear a great context to apply this typology. But before presenting the typology, we first need to define one inevitable concept when analyzing networks: actor centrality. Second, we build upon two previous typologies: Rowley's (1997), linking stakeholder centrality, influence, and pressures; and Savage et al.'s (1991), which accounts for the "threat potential" Rowley and Moldoveanu (2003) linked to identity-based behaviors, along with "cooperation potential," which we argue relates to interest-based behaviors.

Stakeholder Centrality and Centrality Potential

The number of ties varies greatly from one actor to another within a single network. This realization introduces the concept of centrality, a variable attribute of actors also met in the stakeholder perspective. "Firms whose relationships allow them to occupy a more central place in the strategic networks they are a part of enjoy superior returns because of access to better information and opportunities than those firms that are more peripheral" (Gulati, Nohria, & Zaheer, 2000, p. 212). For Ibarra (1993), centrality represents the informal power provided by a structural position in a network. This definition emanates from the aggregation of three centrality dimensions described by Brass and Burkhardt (1993), based on Freeman (1979). First, centrality "degree" or "strength" refers to the number of ties maintained by an actor in a network. Second, "closeness" refers to the shortest distance or path between two actors and determines the independence of an actor in the access of other actors. Third, "betweenness" refers to an actor's capacity to control access to another actor. "Whereas the closeness measure represents a position for avoiding the control of others, the betweenness measure represents potential control, or the possibility of increasing the dependence of others on the actor" (Brass & Burkhardt, 1993, p. 446).

Other features of the structural position of an actor in a network less used to determine its centrality participate in the influence an actor presents in a network. For example, an actor may occupy a structural hole and bridge or span two clusters. This actor will be name "boundary spanner" or "broker" and enjoys considerable influence. Finally, "multiplexity," the diversity of ties maintained by an actor in a network may constitute a type of centrality, especially when one type of ties disappears (Provan, Fish, & Sydow, 2007, p. 484). Centrality is called "asymmetrical" when it differs between two actors.

Yet network position is not sufficient to create benefits for actors; an entrepreneurial approach remains necessary (Burt, 1992; Dhanaraj & Parkhe, 2006). According to Rowley (1997), stakeholders' centrality must be interpreted in relation with the network density, which determines stakeholder influence and pressure. The denser a stakeholder network, the more constraining for stakeholders' action and the more central a stakeholder in a network, the more it may resist the network pressure depending on network density (Figure 2.5).

		Centrality of the focal organization: High	Centrality of the focal organization: Low
Density of the stakeholder network	High	**Compromizer** Negotiates with its stakeholders	**Subordinate** Complies with stakeholder expectations
	Low	**Commander** Controls stakeholder behaviors and expectations	**Solitarian** Avoids stakeholder pressures

Figure 2.5 A structural classification of stakeholder influences: Organizational responses to stakeholder pressures (*Source:* Rowley, 1997).

In Rowley's (1997) typology, two attitudes correspond to high-centrality stakeholders—commander and compromiser—and two attitudes characterize low-centrality stakeholders—namely, subordinate and solitarian. Regardless of network density, we argue that the most central stakeholders in a network will voluntarily or involuntarily control stakeholder behaviors and expectations (commanders), while the least central stakeholders—the more peripheral ones—will instead attempt to avoid other stakeholder pressures (solitarians). Intermediate stakeholders will most likely attempt to negotiate (compromisers) or comply (subordinates) with other stakeholder expectations.

As previously mentioned, according to Rowley and Moldoveanu (2003), stakeholders who are too similar will behave in order to differentiate from one another because of the identity threat they represent to one another. When collaborating with each other, these stakeholders will likely adopt a range of attitudes according to the identity threat that they perceive the other represents. This dimension is taken into account with Savage et al.'s typology (1991, Figure 2.6) to which we now turn.

We argue that relational identification among stakeholders of a network contributes to defining their centrality potential. Savage et al. (1991) mobilized stakeholders' potential for cooperation and stakeholders' potential for threat to create their typology. We use this potential to identity threat to explain the behaviors of too-similar stakeholders towards one another, while the cooperation potential relates to similar interest-based behaviors. This way, relational identification defines stakeholders' collaborative potential, while ambivalent relational identification delineates stakeholders' potential for threat and relational disidentification relates to "marginal" stakeholders or Rowley's "solitarian" stakeholders.

Combining both typologies—Rowley (1997) and Savage et al. (1991)—accounts for some of the complexity encountered in any stakeholder networks. Nevertheless, the following typology refers more specifically to technological innovation networks such as a research consortium encompassing a diversity of stakeholders—industrial, academic, and governmental—typical of high-tech industries.

		Stakeholder's potential for threat to organization	
		High	Low
Stakeholder's potential for cooperation with organization	High	Mixed blessing	Supportive
	Low	Nonsupportive	Marginal

Figure 2.6 Diagnostic Typology of Organizational Stakeholders (*Source:* Savage et al., 1991).

A Typology of Stakeholder Centrality Potential

In line with the stakeholder perspective, some stakeholders are more central than others (more peripheral) in networks, while others occupy intermediate positions. This will likely be the case in most technological innovation networks underlying a supply chain where multinationals granting contracts are central in the network and on top of the supply chain while a constellation of SMEs are peripheral in the network and at the bottom of the supply chain. Other actors involved in specific parallel functions such as research and development, marketing, public relations, distribution, consulting, formation, and so on might occupy even more intermediate positions. This would be the case of many industries generating complex high-tech products counting numerous constituents with often global suppliers such as the automobile, aerospace, computer, or electronics industries. New entrants in these networks will most likely start at the periphery of the network and migrate to more central positions. Otherwise, established stakeholders in the network might eventually reevaluate the relevance of their participation and migrate to more peripheral positions, which are less time- and energy-consuming. In any case, more suitable alternative networks are usually available, and stakeholders may also start their own network in which they can be more central players.

According to Sluss and Ashforth's (2007) typology (Figure 2.4), three scenarios are possible in relational identification: (a) relational identification, (b) relational disidentification, and (c) a mix of both—ambivalent identification. In Figure 2.7, identification and disidentification delimit two axes, where the junction of simultaneous identification and disidentification to the same stakeholder still defines ambivalent identification. Each axis is divided

Relational disidentification with...

Stakeholder	Peripheral	Intermediate	Central
Central	Network ambitious: Very high centrality potential	Network solidary: High centrality potential	Network opportunists: Moderate centrality potential
Intermediate	Network solidary: High centrality potential	Network opportunists: Moderate centrality potential	Network threatened: Low centrality potential
Peripheral	Network opportunists: Moderate centrality potential	Network threatened: Low centrality potential	Network alienated: Very low centrality potential

Figure 2.7 Typology of stakeholder centrality potential from relational identification.

in three simple categories of stakeholders following their centrality in a network—namely central, intermediate, or peripheral stakeholders.

Five types of centrality potential emerge from the junction of the two axes according to their relational identification to other stakeholders: (a) *Network ambitious* stakeholders relationally identify with central stakeholders and relationally disidentify with peripheral stakeholders, which grants them a very high centrality potential; (b) *network solidarity* stakeholders either relationally identify with intermediate stakeholders and relationally disidentify with peripheral stakeholders or identify with central stakeholders and relationally disidentify with intermediate stakeholders, which confers to them a high centrality potential; (c) *network opportunists* are relationally ambivalent in their identification with at least one of the three types of stakeholders, which allow a moderate centrality potential; (d) *network threatened* stakeholders either relationally disidentify with central stakeholders while relationally identifying with intermediate stakeholders or relationally identify with peripheral stakeholders while relationally disidentifying with intermediate stakeholders, indicating a low centrality potential; and finally, (e) *network alienated* stakeholders relationally disidentify with central stakeholders and relationally identify with peripheral stakeholders, only admitting a very low centrality potential. These categories remain ideal types and the reality will expose a mixture of several centrality potentials. The following discussion describes how different elements of this typology relate to Rowley's (1997) and Savage et al.'s (1991) categories.

DISCUSSION AND CROSS-LEVEL IMPLICATIONS

We now discuss how this new typology contributes to integrate two long-standing typologies along with a range of concepts borrowed from the stakeholder and the network literatures. We do so by first more closely examining each of the five identified attitudes related to distinct network centrality potentials. Then the differences between a network perspective, separating individual and collective actors, and a stakeholder perspective, allowing the study of individuals and collectives alongside, led us to explore the typology when applied to different levels of analysis, only to realize the typology does not appear theoretically sensitive to the level of analysis and may be a useful tool to link several levels of interactions.

First, network ambition relates to relational identification, centrality potential, and collaborative potential and corresponds to Salvage et al.'s (1991) "supportive" stakeholders. A more adequate term was preferred since the relational identification relates only to central stakeholders and not, or to a lesser extent, to intermediate and peripheral stakeholders. These stakeholders will remain in the network and attempt to evolve with it and central stakeholders, although their relationships with other stakeholders may suffer from it. However, in order to do so, network-ambitious stakeholders need to maintain the role they identify with in their relation with central stakeholders and experience some reciprocity from them.

Second, network opportunism relates to ambivalent relational identification with specific types of stakeholder and may be associated with Salvage et al.'s (1991) "mixed-blessing" stakeholders. A new term was preferred to previous ones, however, since it involves some kind of temporary compromise despite the ambivalence occasioned by identity threats from too-similar partners, most likely involving an interest-based opportunity. Network opportunists will not remain active members of the network when their interests disappear, unless identification with other stakeholders retains them.

Third, network solidarity also relates to Salvage et al.'s (1991) "mixed-blessing" stakeholders, but this time directed towards the entire network and not towards specific stakeholders, while network threat relates to "nonsupportive" stakeholders. Solidarity remains to the stakeholder network despite disidentifications with peripheral stakeholders, so chances are these stakeholders will remain active members of the network and continue to seek collaborations with other members. Identity threat from and disidentification with central stakeholders, on the other hand, seems more problematic, and threatened stakeholders might linger in the network only to nurture their relational identification with peripheral or intermediate stakeholders.

Network alienation relates to relational disidentification, "solitarian" (Rowley, 1997) and "marginal" (Salvage et al., 1991) stakeholders. This denomination was preferred over the two others since stakeholders who are only identified with peripheral stakeholders, hence disidentified with central ones, are not newcomers and must have passed through events that led them to the current alienated state, either from the (in)actions of more central stakeholders or by having alienated themselves consciously or unconsciously. Network-alienated stakeholders will soon exit the network since their involvement does not justify extensive resource allocation in order to remain active members.

In 2012, Hogg et al. extrapolated the personal relational identity to the collective level in defining an intergroup relational identity. Intergroup relational identity opens up conceptualizations of organizational relational identity as well as relational identity between members from different organizations. This is what we put forward in this chapter.

First, let's consider the interaction between two organizational representatives from two different organizations. Representative A comes from Organization A and Representative B comes from Organization B. Organizations A and B fulfill an organizational role for each other, whereas Representatives A and B fulfill a personal role for one another. An organization may identify or disidentify with the other, and a representative may identify

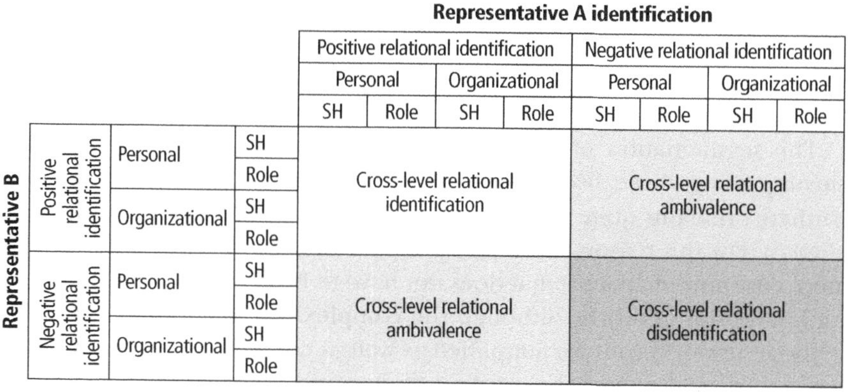

Figure 2.8 Cross-level relational identification.

or disidentify with the other. Furthermore, an organization may identify or disidentify with the organizational role it plays for the other and a representative may identify or disidentify with the personal role he or she plays for the other. Figure 2.8 presents the range of cross-level identification possibilities from the interaction of two representatives attached to different organizations. Luckily, only the three same cross-level scenarios remain possible: cross-level relational identification, cross-level relational disidentification, or cross-level relational ambivalence. Consequently, no matter how many levels of analysis are considered, only three scenarios will emerge that point toward the cross-level analytical qualities of our typology.

Limitations, Implication for Policy, and Directions for Future Research

When a stakeholder identifies with another, it might attempt to imitate and model the behaviors of the identification target. However, from the point of view of the stakeholder being imitated, imitators might easily appear threatening, especially if they are comparable in size. This typology does not fully capture the complex interactive process involved in the stakeholder centrality potential. This is especially true for small players among them since they are present in greater numbers, whereas big corporations (original equipment manufacturers) competing with each other will most likely be part of distinct industrial networks, often as orchestrators. On the contrary, the few big corporations evolving in the same ecosystem or industrial network have usually slowly distinguished themselves with time, will attempt to remain distinct, and work in synergies and complementarity. The story differs for small players highly depending on contracts granted by bigger firms for their survival. Their requirements put pressure on them, and remaining distinct from the competition appears a greater challenge since it appears far more convenient for big corporations to compare and select their suppliers only according to costs.

The segmentation of stakeholders according to their centrality, only encompassing three broad ideal-type categories, although conveniently synthetic, remains quite abstract to fully grasp most networks it may be applied to. For this reason, this typology might be better used as a starter to a more customized analysis that does not have to be restricted to technological innovation networks, although the complexity and constant evolution of these networks calls for simplified as well as complex typologies. Future research should pursue the creation of dynamic and useful typologies, exploring other possible cross-fertilization between the stakeholder and the network approaches and cross-level integrating concepts. As such, identity is a proven syncretic concept detaining the potential to explain individual and collective behaviors in the most abstract processual contexts and should no longer be limited to being a modest antecedent to innovation.

REFERENCES

Albert, S., & Whetten, D. A. (1985). Organizational identity. In L. L. Cummings & B. M. Staw (Eds.), *Research in organizational behavior* (Vol. 7, pp. 263–295). Greenwich, CT: JAI Press.

Alvesson, M. (1990). Organization: From substance to image? *Organization Studies, 11*(3), 373–394.

Alvesson, M., Ashcraft, K. L., & Thomas, R. (2008). Identity matters: Reflections on the construction of identity scholarship in organization studies. *Organization, 15*(1), 5–28.

Ashforth, B. E., Harrison, S. H., & Corley, K. G. (2008). Identification in organizations: An examination of four fundamental questions. *Journal of Management, 34*(3), 325–374.

Ashforth, B. E., & Sluss, D. M. (2006). Relational identities in organizations: Healthy versus unhealthy. In O. Kyriakidou & M. F. Özbilgin (Eds.), *Relational perspectives in organizational studies: A research companion* (pp. 8–27). Cheltenham: Edward Elgar.

Borgatti, S. P., & Foster, P. C. (2003). The network paradigm in organizational research: A review and typology. *Journal of Management, 29*(6), 991–1013.

Brass, D. J., & Burkhardt, M. E. (1993). Potential power and power use: An investigation of structure and behavior. *Academy of Management Journal, 36*(3), 441–470.

Brass, D. J., Galaskiewicz, J., Greve, H. R., & Tsai, W. (2004). Taking Stock of Networks and Organizations: A Multilevel Perspective. *Academy of Management Journal, 47*(6), 795–817.

Brewer, M. B., & Gardner, W. (1996). Who is this "we"? Levels of collective identity and self-representations. *Journal of Personality and Social Psychology, 71*(1), 83–93.

Brickson, S. (2000). The impact of identity orientation on individual and organizational outcomes in demographically diverse settings. *Academy of Management Review, 25*(1), 82–101.

Brickson, S. (2005). Organizational identity orientation: Forging a link between organizational identity and organizations' relations with stakeholders. *Administrative Science Quarterly, 50*(4), 576–609.

Brown, A. D. (2001). Organization studies and identity: Towards a research agenda. *Human Relations, 54*(1), 113–121.

Burt, R. S. (1992). *Structural holes: The social structure of competition.* Cambridge, MA: Harvard University Press.

Capaldo, A. (2007). Network structure and innovation: The leveraging of a dual network as a distinctive relational capability. *Strategic Management Journal, 28,* 585–608.

Chen, S., & Andersen, S. M. (2008). The relational self in transference: Intrapersonal and interpersonal consequences in everyday social life. In J. V. Wood, A. Tesser, & J. G. Holmes (Eds.), *The self and social relationships* (pp. 231–254). New York, NY: Psychology Press.

Choi, J., & Wang, H. (2009). Stakeholder Relations and the Persistence of Corporate Financial Performance. *Strategic Management Journal, 30,* 895–907.

Clarkson, M. B. E. (1995). A stakeholder framework for analysing and evaluating corporate social performance. *Academy of Management Review, 20*(1), 92–117.

Dhanaraj, C., & Parkhe, A. (2006). Orchestrating innovation networks. *The Academy of Management Review, 31*(3), 659–669.

Dutton, J. E., & Dukerich, J. M. (1991). Keeping an Eye on the Mirror: Image and Identity in Organizational Adaptation. *Academy of Management Journal, 34*(3), 517–554.

Dutton, J. E., Dukerich, J. M., & Harquail, C. V. (1994). Organizational image and member identification. *Administrative Science Quarterly, 39,* 239–263.

Eesley, C., & Lenox, M. J. (2006). Firm response to secondary stakeholder action. *Strategic Management Journal,* 27(8), 765–781.

Fanelli, A., & Misangyi, V. F. (2006). Bringing out charisma: CEO charisma and external stakeholders. *Academy of Management Review, 31*(4), 1049–1061.

Fassin, Y. (2009). The stakeholder model refined. *Journal of Business Ethics, 84,* 113–135.

Freeman, L. C. (1979). Centrality in social networks conceptual clarification. *Social Networks, 1,* 215–239.

Freeman, R. E. (1984). *Strategic management: A stakeholder approach.* Boston, MA: Pitman.

Friedman, A., & Miles, S. (2002). Developing stakeholder theory. *Journal of Management Studies, 39*(1), 1–21.

Frooman, J. (1999). Stakeholder influence strategies. *Academy of Management Review, 24*(2), 191–205.

Gioia, D. A., Schultz, M., & Corley, K. G. (2000). Organizational identity, image, and adaptive instability. *Academy of Management Review, 25*(1), 63–81.

Gioia, D. A., & Thomas, J. B. (1996). Identity, image, and issue interpretation: Sensemaking during strategic change in academia. *Administrative Science Quarterly, 41*(3), 370–403.

Gulati, R., Nohria, N., & Zaheer, A. (2000). Strategic networks. *Strategic Management Journal, 21*, 203–215.

Harrison, J., & Freeman, R. E. (1999). Stakeholder, social responsibility, and performance: Empirical evidence and theoretical perspectives. *Academy of Management Journal, 42*(5), 479–485.

Hart, S. L., & Sharma, S. (2004). Engaging fringe stakeholders for competitive imagination. *Academy of Management Executive, 18*(1), 7–18.

Hatch, M. J., & Schultz, M. (1997). Relations between organizational culture, identity and image. *European Journal of Marketing, 31*(5/6), 356–365.

Hatch, M. J., & Schultz, M. (2002). The dynamics of organizational identity. *Human Relations, 55*(8), 999–1018.

Heath, J., & Norman, W. (2004). Stakeholder theory, corporate governance and public management: What can the history of state-run enterprises teach us in the post-Enron era? *Journal of Business Ethics, 53*, 247–265.

Henriques, I., & Sadorsky, P. (1999). The relationship between environmental commitment and managerial perceptions of stakeholder importance. *Academy of Management Journal, 42*(1), 87–99.

Hogg, M. A., Knippenberg, D. V., & Rast, D. E. (2012). Intergroup leadership in organizations: Leading across groups and organizational boundaries. *Academy of Management Review, 37*(2), 232–255.

Ibarra, H. (1993). Network centrality, power, and innovation involvement: Determinants of technical and administrative roles. *Academy of Management Journal, 36*(3), 471–501.

Illia, L., & Lurati, F. (2006). Stakeholder perspectives on organizational identity: Searching for a relationship approach. *Corporate Reputation Review, 8*(4), 293–304.

Jawahar, I. M., & McLaughlin, G. L. (2001). Toward a descriptive stakeholder theory: An organizational life cycle approach. *Academy of Management Review, 26*(3), 397–414.

Jensen, T., & Sandström, J. (2011). Stakeholder theory and globalization: The challenges of power and responsibility. *Organization Studies, 32*(4), 473–488.

Jones, T. M., Felps, W., & Bigley, G. A. (2007). Ethical theory and stakeholder-related decisions: The role of stakeholder culture. *Academy of Management Review, 32*(1), 137–155.

Kacperczyk, A. (2009). With greater power comes greater responsibility? Takeover protection and corporate attention to stakeholders. *Strategic Management Journal, 30*, 261–285.

Laplume, A. O., Sonpar, K., & Litz, R. A. (2008). Stakeholder theory: Reviewing a theory that moves us. *Journal of Management, 34*(6), 1152–1189.

Lawrence, A. T. (2010). Managing disputes with nonmarket stakeholders: Wage a fight, withdraw, wait, or work it out. *California Management Review, 53*(1), 90–113.

Luk, C.-L., Yau, O. H. M., Tse, A. C. B., Sin, L. Y. M., & Chow, R. P. M. (2005). Stakeholder orientation and business performance: The case of service companies in China. *Journal of International Marketing, 13*(1), 89–110.

Mainardes, E. W., Alves, H., & Raposo, M. (2011). Stakeholder theory: Issues to resolve. *Management Decision, 49*(2), 226–252.

Mitchell, R. K., Agle, B. R., & Wood, D. J. (1997). Toward a theory of stakeholder salience: Defining the principle of who and what really counts. *Academy of Management Review, 22*, 853–886.

Neville, B. A., & Menguc, B. (2006). Stakeholder multiplicity: Toward an understanding of the interactions between stakeholders. *Journal of Business Ethics, 66*, 377–391.

Post, J. E., Preston, L. E., & Sachs, S. (2002). Managing the extended enterprise: The new stakeholder view. *California Management Review, 45*(1), 6–28.

Pratt, M. G., & Foreman, P. O. (2000). The beauty of and barriers to organizational theory of identity. *Academy of Management Review, 25*(1), 141–152.

Prentice, D. A. (2001). The individual self, relational self, and collective self. In C. Sedikides & M. B. Brewer (Eds.), *Individual self, relational self, collective self* (pp. 315–326). Philadelphia, PA: Psychology Press.

Provan, K. G., Fish, A., & Sydow, J. (2007). Interorganizational networks at the network level: A review of the empirical literature on whole networks. *Journal of Management, 33*(3), 479–516.

Randel, A. E., Jaussi, K. S., & Standifird, S. S. (2009). Organizational responses to negative evaluation by external stakeholders: The role of organizational identity. *Business & Society, 48*(4), 438–466.

Rowley, T. J. (1997). Moving beyond dyadic ties: A network theory of stakeholder influences. *Academy of Management Review, 22*(4), 887–910.

Rowley, T. J., & Moldoveanu, M. (2003). When will stakeholder groups act? An interest- and identity-based model of stakeholder group mobilization. *Academy of Management Review, 28*(2), 204–219.

Savage, G. T., Nix, T. W., Whitehead, C. J., & Blair, J. D. (1991). Strategies for assessing and managing organizational stakeholders. *Academy of Management Executive, 5*(2), 61–75.

Scott, S. G., & Lane, V. R. (2000). A Stakeholder approach to organizational identity. *Academy of Management Review, 25*(1), 43–62.

Sluss, D. M., & Ashforth, B. E. (2007). Relational identity and identification: Defining ourselves through work relationships. *Academy of Management Review, 32*(1), 9–32.

Sluss, D. M., & Ashforth, B. E. (2008). How relational and organizational identity converge: Processes and conditions. *Organization Science, 19*(6), 807–823.

Sluss, D. M., Ployhart, R. E., Cobb, M. G., & Ashforth, B. E. (2012). Generalizing newcomers' relational and organizational identifications: Processes and prototypicality. *Academy of Management Journal, 55*(4), 949–975.

Tice, D. M., & Baumeister, R. F. (2001). The primacy of the interpersonal self. In C. Sedikides & M. B. Brewer (Eds.), *Individual self, relational self, collective self* (pp. 71–88). Philadelphia, PA: Psychology Press.

Zhang, Y., & Huxham, C. (2009). Identity construction and trust building in developing international collaborations. *Journal of Applied Behavioral Science, 45*(2), 186–211.

CHAPTER 3

DISENTANGLING THE INTELLECTUAL STRUCTURE OF INNOVATION AND M&A LITERATURE

Elona Marku, Manuel Castriotta, and Maria Chiara Di Guardo
University of Cagliari

ABSTRACT

In an ever more globalized and dynamic economy, a firm's capability of generating innovations is crucial for keeping up with the pace of technology and even for surviving in competitive markets. The increasing demand for new technical knowledge, expertise, and capabilities may lead firms to choose fast external technology sourcing such as mergers and acquisitions (M&A) instead of developing uncertain time-consuming in-house innovations.

Based on co-citation analysis, this chapter aims to provide an objective overview of the intellectual structure of the extant innovation studies that examined the multifaceted phenomenon of M&A. More specifically, focusing on 93 highly co-cited academic articles published in top journals between 1990 and 2015, we attempt to identify the most active areas and focal streams

Technological Innovation Networks, pages 47–73

and to suggest potential future trends. Five main themes—M&A innovation performance, technology-driven M&A, M&A organizational adaption, M&A integration process, multinational M&A—and several subfields are analyzed and discussed more in depth.

In an ever more globalized and dynamic economy, a firm's capability of generating innovations is crucial for keeping up with the pace of technology and even for surviving in competitive markets (Crossan & Apaydin, 2010; Tushman & O'Reilly, 1996). The increasing demand for novel technologies fosters the need to acquire rapidly new technical knowledge, expertise, and capabilities, essential for the whole inventive activity process (Valentini & Di Guardo, 2012). This situation may lead firms to choose external technology sourcing such as mergers and acquisitions (M&A) to avoid uncertain, time-consuming, or path-dependent organic deficiencies (Higgins & Rodriguez, 2006). The great complexity of this phenomenon, despite decades of intensive study, continues to attract scholars from different research fields, ever increasing the overall quantity of literature (Cartwright & Schoenberg, 2006; Di Guardo & Valentini, 2007; Valentini & Di Guardo, 2012).

According to Gopalakrishnan and Damanpour (1997), better knowledge of the structure of a body of literature leads to a more accurate understanding of the value of research in each of its streams. For this reason, this chapter aims to provide an overview of the intellectual structure of the extant innovation studies that examined the complex phenomenon of M&A. To be clear, while neither the importance nor the relevance of both innovation and M&A literature are not questioned, the plethora of different interpretations highlights the need to disentangle M&A's multifaceted dynamics within the innovation arena.

Among existing innovation literature reviews, Wolfe (1994) presented a conceptual review of the organizational innovation literature and distinguished three main streams according to research question, research approach, and research address. Later, Gopalakrishnan and Damanpour (1997) reviewed innovation research from three different fields (economics, organizational sociology, and technology management) and proposed several dimensions of innovation: level of analysis, stage of innovation process, and type of innovation.

Most of the M&A literature reviews focus on financial and market research (see Agrawal & Jaffe, 2000); however, O'Neill, Pouder, and Buchholtz (1998) provided one of the first contributions that examined the relationship between M&A and the diffusion of innovation. Cartwright and Schoenberg (2006) analyzed a collection of 35 high-quality papers and identified three main streams of M&A research inquiry: strategic fit, organizational fit, and acquisition process. De Man and Duysters' (2005) examination of

15 studies highlighted the importance of M&A's effects on the innovative capabilities of the parties involved, describing a three-directional stream: positive, negative, and even neutral.

In contrast to the studies mentioned above that follow a *qualitative* approach, our study applies a *quantitative* technique based on bibliometric analysis. To achieve our goals, we examine the 93 most influential strategy research papers published in top journals from 1990 to 2015 that treated the themes of M&A within the academic area of innovation. Our study is based on the co-citation technique (Di Guardo & Harrigan, 2012; Nerur, Rasheed, & Natarajan, 2008; Ponzi, 2002; Ramos-Rodriguez & Ruíz-Navarro, 2004; White & McCain, 1998), which refers to the similarities found between two papers based on third-party papers that cite the first two in correlation with each other. Based on that, we identify relationships and map the different dominant approaches to strategy research that treat the innovation and M&A domains. Following the literature (Di Guardo, Galvagno, & Cabiddu, 2012; McCain, 1990; Nerur et al., 2008), we conduct a multivariate analysis and finally interpret our findings.

METHODOLOGY

Overview of the Method

This literature review explores the intellectual structure of the innovation and M&A research field in order to better understand its origins, current state of development, and future trends. In so doing, we use the co-citation analysis as a validated bibliometric technique not only to describe and condense a body of literature, but also to display, map, and interpret it, despite the increasing knowledge complexity (Boyack & Klavans, 2010; Garfield, 1979; Nerur et al., 2008; Ponzi, 2002; Ramos-Rodriguez & Ruíz-Navarro, 2004; Small, 1973,1977; White & McCain, 1998).

Co-citation is based on a count of the number of times two studies are jointly cited in subsequent contributions (Crane, 1972; Nerur et al., 2008; Small, 2003, 2010). By extension, co-citation can be represented by a frequency matrix of co-occurrences, in which a larger unit of analysis consisting of n items is paired by citing articles (Small, 1973). A high frequency of co-citations signals a strong relationship between the jointly cited contributions, in this way making it more probable that co-cited papers belong to the same school of thought and social and conceptual structure: the so called *invisible college* (Crane, 1972; Small, 1977; White & McCain, 1998).

Sample Selection

Our empirical analysis partially follows the McCain (1990) prescriptions that deal with selection of the unit of analysis and its forward citations, filtering of the unit of analysis based on citation counts, generation of co-citation frequencies, filling of a raw citation matrix and its conversion into a VOSviewer algorithm correlation matrix, multivariate analysis, and interpretation of results.

From a methodological point of view, this study attempts to achieve a dual goal: (a) extraction of a representative sample and (b) identification of the largest number of contributions in "innovation" that focus on "M&A." In so doing, several choices are needed to identify the keywords, database, research areas, and unit of analysis.

We selected two contributions that helped us identify the keywords for the research areas of interest. On one hand, Di Guardo and Harrigan (2012) provided us keywords related to innovation topics: innovation, R&D, and high technology; on the other hand, Santos, Basso, Kimura, and Kayo (2014) provided us with the following keywords related to M&A literature: "mergers & acquisitions, mergers and acquisitions, M&A, mergers, acquisitions, and consolidation & merger of corporations." These studies allowed us to fix two anchor points.

The choice of database is crucial, as each *container* follows heterogeneous protocols and complementary logic that reverberate in the results of science mapping analysis (Loi, Castriotta, & Di Guardo, 2016). The source of our empirical data is the ISI Web of Science database, which was selected for several reasons: (a) it adopts an entirely quantitative protocol in evaluating "scientific" contributions primarily based on impact factor index (IF); (b) it is the most widely adopted source in the literature, making its academic contributions effectively comparable.

This study covers a 25-year timespan, focusing on the period 1990–2015. This period represents the growth phase of M&A research, and the wide interval in part overcomes some limitations of co-citation analysis, as a certain amount of time is needed for a contribution to be cited (Castriotta & Di Guardo, 2016). In line with the objective of research, we focus our attention on the areas of business, management, and economics.

The selected keywords were searched for the title, abstract, or keywords of each academic contribution. The string used on the ISI Web of Science was: TS= (("mergers") or ("acquisitions") or ("mergers and acquisitions") or ("mergers & acquisitions") or ("M&A*") or ("consolidation* $ merger* of corporation*)). The results from the 1990–2015 period showed 8,508 studies for M&A and 88,395 for innovation research fields. By using the Boolean operator "and," we identified the articles that deal with both innovation and M&A topics. We then filtered for the three areas of business, management, and economics, and excluded other document types such as conference proceedings, editorials, and so on. This procedure reduced our sample to 514 elements.

To balance our interest in defining the core of the discipline with our need to increase the breadth of citations that characterizes our unit of analysis, we therefore adopted an *ad hoc* selection criterion for retrieving co-citation frequency (Béchard & Grégoire, 2005; Small, 1977; Stern, 2014). Specifically, for papers issued from 1991–2009, 30 citations were necessary to be included in the unit of analysis. For articles published from 2010–2012, criteria stipulated at least 20 citations or at least two citations on average per year, while articles published from 2013–2015 timespan required at least 10 citations. Following these criteria, we selected 99 articles among the top 204 most cited articles, including 81 with at least 50 citations, 13 with at least 30 citations, and 6 with at least 10 citations. At this point, in following recent bibliometric research and to overcome some of the artifacts of multidimensional scaling in mapping co-citation data, we used an alternative technique: the VOSviewer computer program algorithms (Appio, Cesaroni, & Di Minin, 2014; van Eck & Waltman, 2007, 2009; van Eck, Waltman, Den Berg, & Kaymak, 2006; van Eck, Waltman, Dekker, & van den Berg, 2010; Waaijer, van Bochove, & van Eck, 2010).

From analysis of the similarities between the 99 items through automatic generation of a co-occurrence matrix, it emerged that 6 academic contributions had no connection with each other. For this reason, the unit of analysis was definitively reduced to 93 items, as shown in Table 3.1. Although these contributions represent less than 20% of the retrieved papers, their weight accounts for 64% of the total number of citations of 514 articles, making our sample highly representative for investigation.

Finally, five main groups and the main links between contributions were detected through adoption of a weighted and parameterized variant of modularity-based clustering and network analysis, as shown in Table 3.2 (Newman & Girvan, 2004; van Eck et al., 2010).

DISCUSSION OF RESULTS

Discussion of Results: Citation Analysis

Figure 3.1 details the 32 out of 38 journals that treated the theme of M&A within the innovation research field. The long tail demonstrates the presence of many publishing journals that appear only once in the whole sample; however, many of these articles are characterized as having a strong impact and influence in terms of citations. This is the case of the most cited article, Gupta, Smith, and Shalley (2006), published in *Research in Organizational Behavior.*

TABLE 3.1 The Set of 93 Articles Ordered by Frequency of Citations (from most- to least-cited)

P1: Gupta et al. (2006)	P32: Hoskisson, Hitt, & Hill (1993)	P63: Sher and Yang (2005)
P2: Makadok (2001)	P33: Allen and Pantzalis (1996)	P64: Belderbos (2003)
P3: Ahuja and Katila (2001)	P34: Wang and Zajac (2007)	P65: Blonigen and Taylor (2000)
P4: Haunschild and Miner (1997)	P35: Zahra (1995)	P66: Rothaermel and Thursby (2007)
P5: Hitt et al. (1996)	P36: O'Neill et al. (1998)	P67: Paruchuri et al. (2006)
P6: Sitkin (1992)	P37: De Man and Duysters (2005)	P68: Kelm et al. (1995)
P7: Healy et al. (1992)	P38: Eisenbeiss et al. (2008)	P69: Ailon-Souday and Kunda (2003)
P8: Morosini et al. (1998)	P39: Higgins and Rodriguez (2006)	P70: Eun et al. (1996)
P9: Bresman et al. (1999)	P40: Cassiman et al. (2005)	P71: Ofek and Sarvary (2001)
P10: Hagedoorn and Cloodt (2003)	P41: Prabhu et al. (2005)	P72: Ernst and Vitt (2000)
P11: Raisch et al. (2009)	P42: Kitchener (2002)	P73: Porrini (2004)
P12: King et al. (2004)	P43: Graves and Langowitz (1993)	P74: Vaara and Monin (2010)
P13: Frost et al. (2002)	P44: Cloodt et al. (2006)	P75: Duysters and De Man (2003)
P14: Hoskisson, Hitt, Johnson, & Moesel, 1993	P45: Puranam and Srikanth (2007)	P76: Salvato et al. (2010)
P15: Hayward (2002)	P46: van de Vrande et al. (2006)	P77: Nemanich and Vera (2009)
P16: Capron et al. (1998)	P47: Buckley and Casson (1996)	P78: van de Vrande et al. (2009)
P17: Ranft and Lord (2002)	P48: Makri et al. (2010)	P79: Aybar and Ficici (2009)
P18: Karim and Mitchell (2000)	P49: Brau et al. (2003)	P80: Li (2009)
P19: Hambrick and Cannella (1993)	P50: Håkanson and Nobel (1993)	P81: Faleye et al. (2011)
P20: Chi (1994)	P51: Lengnick-Hall (1992)	P82: Flatten et al. (2011)
P21: Fichman and Kemerer (1999)	P52: Frost and Zhou (2005)	P83: Officer et al. (2009)
P22: Hitt et al. (1994)	P53: Marsh and Stock (2003)	P84: Reitzig and Puranam (2009)
P23: Pennings et al. (1994)	P54: Achilladelis and Antonakis (2001)	P85: Benson and Ziedonis (2009)
P24: Puranam et al. (2006)	P55: Marsh and Stock (2006)	P86: Curran and Leker (2011)
P25: Graebner (2004)	P56: Groysberg et al. (2008)	P87: Zollo and Reuer (2010)
P26: Vassolo et al. (2004)	P57: Goodman and Dion (2001)	P88: O'Reilly et al. (2009)
P27: Lambe and Spekman (1997)	P58: Chaudhuri and Tabrizi (1999)	P89: Hagedoorn and Wang (2012)
P28: Vanhaverbeke et al. (2002)	P59: Zahra and Hayton (2008)	P90: Bresman (2013)
P29: Stuart and Sorenson (2003)	P60: Long and Ravenscraft (1993)	P91: Seru (2014)
P30: Deng (2009)	P61: Tsang and Yip (2007)	P92: Han et al. (2012)
P31: Gambardella and Torrisi (1998)	P62: Karim (2006)	P93: Fischer and Henkel (2012)

TABLE 3.2 The Research Groups After the Cluster Analysis

M&A Innovation Performance	Technology-Driven M&A	Organizational Adaption	M&A Integration Process	Multinational M&A
P3: Ahuja & Katila (2001)	P10: Hagedoorn & Cloodt (2003)	P1: Gupta et al. (2006)	P8: Morosini et al. (1998)	P4: Haunschild & Miner (1997)
P5: Hitt et al. (1996)	P27: Lambe & Spekman (1997)	P2:Makadok (2001)	P9: Bresman et al. (1999)	P6: Sitkin (1992)
P7: Healy et al. (1992)	P28: Vanhaverbeke et al. (2002)	P11: Raisch et al. (2009)	P12: King et al. (2004)	P13: Frost et al. (2002)
P14: Hoskisson, Hitt, Johnson, & Moesel, 1993	P31: Gambardella & Torrisi (1998)	P18: Karim & Mitchell (2000)	P16: Capron et al. (1998)	P15: Hayward (2002)
P22: Hitt et al. (1994)	P37: De Man & Duysters (2005)	P38: Eisenbeiss et al. (2008)	P17: Ranft & Lord (2002)	P20: Chi (1994)
P23: Pennings et al. (1994)	P39: Higgins & Rodriguez (2006)	P53: Marsh & Stock (2003)	P19: Hambrick & Cannella (1993)	P21: Fichman & Kemerer (1999)
P32: Hoskisson, Hitt, & Hill, 1993	P40: Cassiman et al. (2005)	P55: Marsh & Stock (2006)	P24: Puranam et al. (2006)	P26: Vassolo et al. (2004)
P35: Zahra (1995)	P44: Cloodt et al. (2006)	P58: Chaudhuri & Tabrizi (1999)	P25: Graebner (2004)	P30: Deng (2009)
P43: Graves & Langowitz (1993)	P46: van de Vrande et al. (2006)	P62: Karim (2006)	P29: Stuart & Sorenson (2003)	P33: Allen & Pantzalis (1996)
P54: Achilladelis & Antonakis (2001)	P65: Blonigen & Taylor (2000)	P71: Ofek & Sarvary (2001)	P41: Prabhu et al. (2005)	P34: Wang & Zajac (2007)
P59: Zahra & Hayton (2008)		P77: Nemanich & Vera (2009)	P42: Kitchener (2002)	P36: O'Neill et al. (1998)
P60: Long & Ravenscraft (1993)		P88: O'Reilly et al. (2009)	P45: Puranam & Srikanth (2007)	P47: Buckley & Casson (1996)
		P89: Hagedoorn & Wang (2012)	P48: Makri et al. (2010)	P50: Håkanson & Nobel (1993)
			P49: Brau et al. (2003)	
			P51: Lengnick-Hall (1992)	

(continued)

TABLE 3.2 The Research Groups After the Cluster Analysis (continued)

M&A Innovation Performance	Technology-Driven M&A	Organizational Adaption	M&A Integration Process	Multinational M&A
P63: Sher & Yang (2005) P68: Kelm et al. (1995) P70: Eun et al. (1996) P82: Flatten et al. (2011) P83: Officer et al. (2009) P91: Seru (2014)	P66: Rothaermel & Thursby (2007) P75: Duysters & De Man (2003) P78: van de Vrande et al. (2009) P80: Li (2009) P81: Faleye et al. (2011) P85: Benson & Ziedonis (2009) P86: Curran & Leker (2011) P92: Han et al. (2012) P93: Fischer & Henkel (2012)		P57: Goodman & Dion (2001) P67: Paruchuri et al. (2006) P69: Ailon-Souday & Kunda (2003) P72: Ernst & Vitt (2000) P74: Vaara & Monin (2010) P76: Salvato et al. (2010) P84: Reitzig & Puranam (2009)	P52: Frost & Zhou (2005) P59: Groysberg et al. (2008) P61: Tsang & Yip (2007) P64: Belderbos (2003) P73: Porrini (2004) P79: Aybar & Ficici (2009) P87: Zollo & Reuer (2010) P90: Bresman (2013)

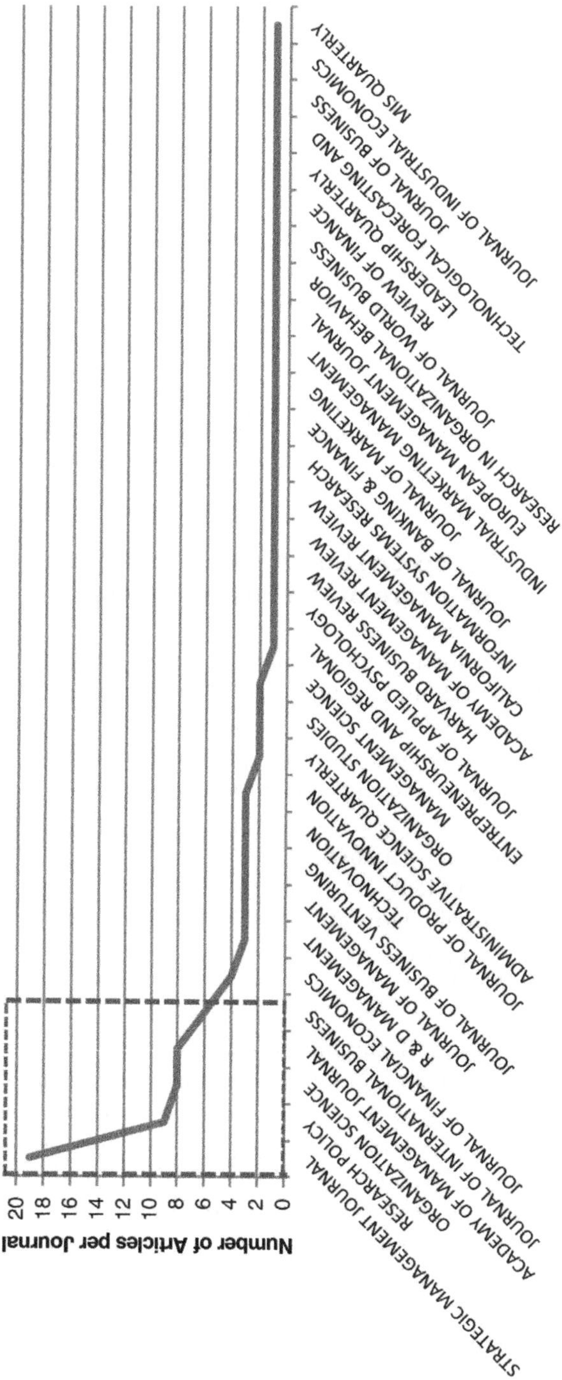

Figure 3.1 Number of articles per journal.

The further inspection of the papers journal-wise reveals various interesting points. First, the key player in this research arena is *Strategic Management Journal*, one out of every five most important works has been published here. The dashed rectangle (Figure 3.1) highlights the area including half of the contributions examined, which are concentrated on five top journals: *Strategic Management Journal, Research Policy, Academy of Management Journal, Organization Science,* and *Journal of International Business Studies.* These papers received more than 62% of the total citations. Second, empirical works and quantitative methods strongly prevail with respect to more theoretical ones: only one article has been published in *Academy of Management Review.* Last, although most of the studies are published in journals with a more pronounced academic cut, with emphasis on analysis rather than on normative prescription, there are three papers published on practitioner-oriented journals such as *Harvard Business Review, Leadership Quarterly,* and *California Management Review.*

Discussion of Results: Co-Citation Analysis and Overview of the Intellectual Structure of the Research Field

The VOSviewer software allows to reproduce the co-citation cluster analysis as shown in Table 3.2, and to map the 93 filtered contributions in a single area. Figure 3.2 illustrates the different angulations from which the M&A topics have been analyzed more deeply within the innovation field. Five core themes emerge using the co-citation technique:

- M&A innovation performance
- Technology-driven M&A
- M&A organizational adaption
- M&A integration process
- Multinational M&A

It should be said that the positioning inside the map is not random, and therefore it can provide meaningful and key information. More specifically, two or more articles belong to the same research group if they share some co-citation features, or if they synthetize a similar appeal that brings other authors to cite them jointly. Of course, this does not mean that they should necessarily share the same identical research questions or even the same findings.

The articles that are limited at the boundaries of the co-citation map, surrounded by fewer neighbors, indicate that their role is not so central in the research arena. Last, the empty areas also signal a differentiation between groups: They indicate the potential gaps in the extant literature and suggest future research streams.

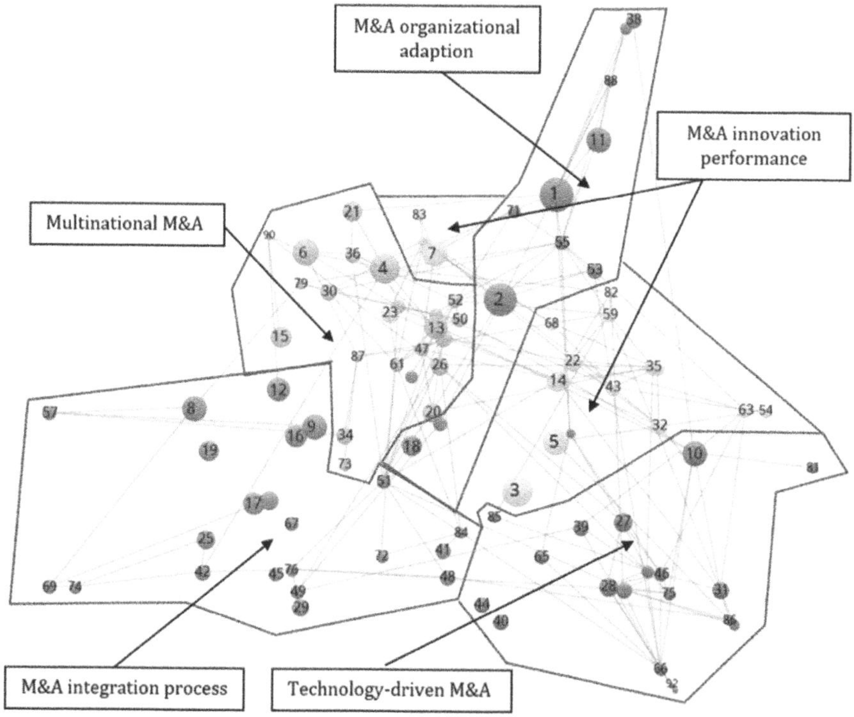

Figure 3.2 The literature mapping and network.

As suggested by the articles' positioning illustrated by Figure 3.2, the multinational research group seems to play a central role for both the innovation and M&A literature. Not surprisingly, this cluster is located in proximity to the M&A integration process group, probably as they share a more economics cut. The M&A integration process and technology-driven M&A groups are located near the boundaries of the map and present no overlap with their respective neighboring clusters. Instead, the organizational adaption group can be considered a transversal research theme: One group of articles is scattered within the multinational M&A and M&A innovation performance groups, while others are positioned in the upper part of the map with few neighbors. The widest empty space is located between the M&A innovation performance research group and the multinational M&A research group, indicating a neglected region in the very core of the mapped research arena.

The network analysis (Figure 3.2) allows identification of the strongest relationships among the unit of analysis and different research groups. Although many of contributions to the M&A integration process group have a strong impact in terms of citations (see Table 3.2), this research area does not otherwise seem to be very central in the innovation and M&A literature. In fact, it presents the lowest link density, suggesting that co-citations often occur between the components of the same cluster. The strongest links deal with the multinational M&A and M&A innovation performance groups. Also, the M&A organizational adaption group is linked with two other groups: the multinational M&A and M&A innovation performance groups. The technology-driven M&A stream appears to be more related to M&A innovation performance and somehow with the M&A integration process groups.

Meanwhile, the M&A innovation performance and multinational M&A research groups can be considered more central themes, as illustrated by their dense and intertwined network. In particular, multinational M&A is strongly connected with M&A performance, M&A integration process, and M&A organizational adaption, but is weakly linked with technology-driven M&A. This is probably due to the level of analysis: The first focus is on the multinational level, while the second is on an industry level. The M&A innovation performance network suggests dense link coverage between the examined clusters, including in its connections to the multinational M&A, M&A organizational adaption, technology-driven M&A, and M&A integration process groups.

M&A Innovation Performance Research Group

This research group consists of 18 academic articles that pay particular attention to the role of M&A innovation performance. Scholars such as Hitt, Hoskisson, Johnson, and Moesel are considered pioneers of the combined innovation and M&A literature thanks to their earlier examination of several M&A dynamics within the innovation field. The overall age of the papers also shows clear evidence of the pioneering patterns: 13 out of 18 articles were published between 1992 and 2001, and the mean publishing of this cluster is the lowest of the sample with a value around 1999.

The contributions of this group highlight three broad research questions that primarily relate to: (a) whether M&A helps or hinders innovation performance, and whether the latter affects the post-deal financial domain; (b) which are the drivers of innovation and how these impact innovation performance after an M&A; and (c) several measurement issues. The predominant approach used in these papers is based on empirical analysis.

The innovation performance, representing the gravitational center of this research group, is analyzed from two different angles: as dependent variable of the M&A event or as an independent variable of the post-deal financial performance. In fact, some scholars seem more interested in observing and measuring the direct effect of the M&A organization *shock* on innovation performance (Achilladelis & Antonakis, 2001; Ahuja & Katila, 2001; Graves & Langovitz, 1993; Hitt, Hoskisson, & Ireland, 1994; Hitt, Hoskisson, Johnson, & Moesel, 1996; Long & Ravenscraft, 1993; Zahra,

1995; Zahra & Hayton, 2008), while others prefer to examine the impact of innovation on post-deal financial performance (Hitt et al., 1994; Sher & Yang, 2005), and yet another group tries to capture both of these aspects (Hitt et al., 1994).

Consistent with previous literature reviews, the articles present contrasting results regarding the directionality of M&A's effects on innovation performance: Some contributions found it to be positive (Ahuja & Katila; 2001; Zahra, 1995), others negative (Hitt et al., 1996; Graves & Langowitz, 1993; Long & Ravenscraft, 1993; Zahra & Hayton, 2008), and others even neutral (Healy, Palepu, & Ruback, 1992). Moreover, innovation performance seems to improve the post-deal financial performance (Sher & Yang, 2005) but not in the long term (Hitt et al., 1996).

Some other works attempt to identify the potential drivers of innovation, focusing on the role of diversification (in its broadest sense) and the moderating effect of the firm's absorptive capacity. More specifically, Hitt et al. (1994) found that international diversification enhances innovation performance, while product diversification affects it negatively. Through a quasi-experiment on failed M&A, Seru (2014) revealed evidence that organizational diversification hinders the achievement of innovation novelty within the firm boundaries. Pennings, Barkema, and Douma (1994) used an organizational learning approach to investigate the causes of the success or failure of corporate diversification. Zahra and Hayton (2008) pointed out the key role of absorptive capacity on R&D performance, suggesting that executives give great attention to internal capabilities rather than external sourcing.

The analysis of this research group is based on different innovation performance measures differentiated between input indicators, such as R&D expenditure, R&D intensity (Ahuja & Katila, 2001; Graves & Langowitz, 1993; Hitt et al., 1996; Healy et al., 1992; Hoskisson, Hitt, & Hill, 1993; Long & Ravenscraft, 1993), and output indicators like patents or new products (Achilladelis & Antonakis, 2001; Ahuja & Katila, 2001; Hitt et al., 1996). However, the need for better assessment emerges, especially when the innovation drivers are examined; see the entropy measure of diversification strategy proposed by Hoskisson, Hitt, Johnson, and Moesel (1993) and the more appropriate measure of absorptive capacity developed and validated by Flatten, Engelen, Zahra, and Brettel (2011).

Technology-Driven M&A Research Group

The technology-driven M&A research group includes 19 highly cited contributions (20% of the whole sample). As suggested by the group's publishing year interval (1997–2012) as well as its publishing year mean (2006), which is the lowest of the sample, this research stream is quite recent. Most of the papers focus on empirical analysis at the industry level.

According to the studies in this group, innovation is a result of the R&D strategy and open innovation. The main research questions of these academic works are concerned with (a) whether external technology sourcing can help firms to gain and sustain a competitive advantage, (b) what are the drivers of technological change, and (c) how to measure the post-deal technological performance.

M&A is considered by most as an effective means of external technology sourcing, and its importance is crucial for the acquisition of new technological knowledge, expertise, and capabilities in order to reduce the organic deficiencies (Higgins & Rodriguez, 2006). Cloodt, Hagedoorn, and Van Kranenburg (2006), distinguishing between technological and non-technological M&A, show evidence of a positive effect of the former on innovation performance, while this relationship resulted negatively in the last case. On the contrary, according to Blonigen and Taylor (2000), the high-technology industries emphasize the negative relationship between M&A and innovation performance in terms of R&D intensity. The results seem to be inconclusive.

Furthermore, there is an attempt to go beyond the direct effects of M&A on innovation performance alone, with a focus on the key drivers of technological dimension, such as technological diversification and relatedness. Gambardella and Torrisi (1998) questioned what kind of relationship exists between technological and business diversification. Access to new and diverse resources is often considered an *active ingredient* that improves the innovation process outcomes and is essential for the generation of radical and breakthrough innovations. However, this relationship can also exhibit a curvilinear shape (Cloodt et al., 2006). Moreover, Cassiman, Colombo, Garrone, and Veugelers (2005, p. 195) showed that "technological and market-relatedness between M&A partners distinctly affect the inputs, outputs, performance and organizational structure of R&D process."

Patents emerge as an effective and reliable measure of the inventive activity performance and importance (Blonigen & Taylor, 2000; Cloodt et al., 2006; Curran & Leker, 2011; Gambardella & Torrisi, 1998; Li, 2009; Rothaermel & Thursby, 2007). The literature provides many patent measures as those for monitoring industrial convergence (Curran & Leker, 2011), but sometimes there is no advantage in using multiple indicators (Hagedoorn & Cloodt, 2003).

This research group seems to have strong potential to investigate further technology-driven M&A's effects on the nature of knowledge in terms of quality of innovation process outputs and outcomes.

Organizational Adaption Research Group

The 13 studies included in the organizational adaption research group are some of the youngest in our sample. Considering the chronological order, the first paper was published in 1999, while the last one was published in 2012. Although scholars' interest in this area is quite recent, the presence of the most influential article, Gupta et al. (2006), with 503 citations, suggests the high potential of this research field.

This group's main themes deal with: (a) the role of exploitation/exploration/ambidexterity in organizational adaption and innovation (also connected to the M&A event) and (b) the role of knowledge, capabilities, and resource recombination for innovation success. The topics covered by this group may seem totally aligned with organization studies; nevertheless, several managerial and practical implications are explored by contributions published in practitioner journals such as *Harvard Business Review*, *Leadership Quarterly*, and *California Management Review*.

The special issues of Gupta et al. (2006) and Raisch, Birkinshaw, Probst, and Tushman (2009) aimed to disentangle the foundations of the organizational adaption literature, highlighting the importance of both exploration and exploitation for an organization's success, including innovation. The former work recognized the need to address matters of definition in the light of March's (1991) seminal work, and to go further by examining the main features of this phenomenon, focusing on continuity vs. orthogonality, ambidexterity vs. punctuated equilibrium, and duality vs. specialization. According to Raisch et al. (2009), ambidexterity is the outcome of a dynamic process that involves both individual and organizational levels, requires particular attention in management of the tensions between differentiation and integration, and depends on the ability to integrate both internal and external knowledge. The importance of ambidexterity is explored even by the more practitioner-centered work of O'Reilly, Tushman, and Harreld (2009, p. 34), who showed how "organizations that are able to both explore and exploit are more likely to adapt than organizations that can do only one or the other."

"The first step to a successful acquisition is understanding what capabilities you really need" (Chaudhuri & Tabrizi, 1999, p. 125), because both knowledge and capabilities are essential to fostering innovation and gaining competitive advantage. These elements can be built internally through organic innovation, or they can be acquired externally through M&A or alliances. The latter are found to be both complementary and substitutes for one another, according to the levels of in-house R&D investment (Hagedoorn & Wang, 2012). However, increased acquisitions (e.g., resource-picking) will lead to decreased internal R&D (Makadok, 2001). In addition, knowledge retention enhances product innovation and long-term competitive advantage, but in the case of M&A may worsen inter-temporal integration issues (Marsh & Stock, 2003, 2006).

Acquisitions are also used to undertake path-breaking changes, thanks to substantial leaps into new sets of skills and capabilities, with the aim of achieving long-term business reconfiguration (Karim, 2006; Karim & Mitchell, 2000).

M&A Integration Process Research Group

This research group consists of 22 articles that investigate the multifaceted dynamics of the M&A integration process. The papers are not very recent; most of them are concentrated in the last decade (16 out of 22 were published between 2001 and 2010). The subfields deal with (a) a deeper focus at the firm level and (b) the exploration of more complex cross-border acquisitions scenarios (Morosini, Shane, & Singh, 1998).

From a firm-level perspective, the acquisition of new technologies and capabilities is seen as essential for enhancement of the post-deal innovation performance (Ranft & Lord, 2002). However, M&A often generate several integration issues, mainly connected with human resources management and more specifically with executives (e.g., Hambrick & Cannella, 1993) and inventors (e.g., Ernst & Vitt, 2000; Paruchuri, Nerkar, & Hambrick, 2006), which may strongly affect performance directionality. Integration is a crucial means to enable technological knowledge transfer from the target to the acquirer, and to improve the whole innovation process outcome (Puranam & Srikanth, 2007). Effective integration and implementation are fundamental for the success of M&A (Graebner, 2004). Puranam, Singh, and Zollo (2006, p. 263) also pointed out the coordination–autonomy dilemma and how "large established firms acquiring small, technology-based firms must manage them so as to both exploit their capabilities and technologies in a coordinated way and foster their exploration capacity by preserving their autonomy." In addition, as the role of the acquired executives is instrumental for capturing expected and serendipitous values and further mitigating potential conflicts between autonomy and integration (Graebner, 2004), Hambrick and Cannella (1993) questioned whether their departure rates after an M&A are due to changes in their autonomy that may generate an acrimonious climate.

The macro-level analysis of the articles in this research group highlights the central role of the national cultural distance, intended as the set of routines and repertoires, that positively affect post-deal innovation performance (Morosini et al.,1998). In addition, the study of Bresman, Birkinshaw, and Nobel (1999) explored changes in patterns of knowledge over time, suggesting that tacit knowledge is hard to be transfer rapidly.

Multinational M&A

The 21 articles comprising this research group present a marked international and multinational slant with a constant presence in the innovation and M&A arena. The publishing year interval spans from 1992 to 2013 with a highly cited publication nearly every year.

Multinational M&A research studies adopt a slightly theoretical approach, and indeed here we find the only contribution published in the *Academy of Management Review*. Several lenses are used to observe M&A phenomena: for example, economic theory (Buckley & Casson, 1996), organizational learning perspective (Bresman, 2013; Hayward, 2002; Haunchild & Miner, 1997; Porrini, 2004; Tsang & Yip, 2007; Zollo & Reuer, 2010), institutional theory (Deng, 2009; O'Neill et al., 1998), neoinstitutional theory (Haunchild & Miner, 1997), transaction costs theory (Belderbos, 2003; Chi, 1994), and resource-based view (Chi, 1994; Wang & Zajac, 2007).

M&A are an effective means of entering new markets and spreading innovation (O'Neill et al., 1998). The main subfields that are observed concern (a) M&A as an entry mode that affects innovation performance and (b) the role of past experience.

Many studies have analyzed the determinants that underlie the choice between M&A and alliances, highlighting the role of acquisitions as an important instrument for both trading and combining strategic resources (Buckley & Casson, 1996; Chi, 1994; Wang & Zajac, 2007). Cross-border M&As in particular are often used to access and source strategic assets from foreign countries, essential to gaining and sustaining the competitive advantage in multinational markets (Belderbos, 2003; Deng, 2009).

Bresman (2013) showed how groups change routines, drawing on the prior related experience of other groups. Furthermore, the role of past experience in the case of an alliance is on the one hand an important determinant that positively affects the post-deal innovation performance (Porrini, 2004), while on the other it may generate negative spillover effects if there are dissimilarities between activities (Zollo & Reuer, 2010). Hayward (2002) questioned *when* firms learn from their acquisition experience, pointing out that M&A performance is positively related to prior experience that doesn't have highly similar or dissimilar characteristics with respect to the focal M&A, and that it is associated with small losses. This latter point is also discussed in the contributions of Haunchild and Miner (1997), who revealed asymmetries between learning from success and failure, and Sitkin (1992), who considered small losses as a strategy of learning through failure.

CONCLUSIONS

This chapter provides an objective overview of the intellectual structure of innovation studies that analyzed the complex and multifaceted phenomenon of mergers and acquisitions (M&A). The work's primary motivation lies in the increasing role of M&A in both innovation strategy and innovation management research, as well as the lack of a systematic examination of the conjoint innovation and M&A fields. We apply the co-citation analysis as a reliable bibliometric technique to identify the most active areas and focal streams of a body of literature, and our investigation focuses on 93 highly co-cited contributions from the last 25 years (1990–2015).

The findings show the quantitative approach's strong predominance versus the qualitative perspective. The positioning of the articles in the co-citation map (Figure 3.1) suggests an initial distinction based on two dimensions. The vertical dimension highlights the level of analysis (from the top to the bottom): micro-level vs. industrial-level topics. Meanwhile, the horizontal dimension refers to the different strategies adopted by firms (from the left to the right): organization strategy vs. R&D strategy.

The field is characterized by several fragmentations, although of not excessive size, enabling us to identify five core themes that are shaping and influencing both innovation and M&A scenarios:

1. The M&A innovation performance group represents pioneer research, which contains the earliest stage of the innovation and M&A literature. The interests of the scholars in this cluster focus on the impact of M&A on the innovation performance, devoting less attention to identification of innovation's main drivers.
2. Studies from the technology-driven M&A research group consider innovation a result of the R&D strategy and open innovation, while M&A is seen as an effective means of external technology sourcing. These works' primary aim is to identify clearly the nature of technological knowledge that affects post-deal innovation performance and consequently to find evidence of the drivers of innovation that lead to M&A success.
3. The M&A organizational adaption represents the more recent and transversal stream, which has practical and managerial implications, suggesting this topic's large potential for future studies. This area's contributions point out the central role of knowledge, capabilities, and resource recombination, and more specifically the effects of exploitation/exploration/ambidexterity on organizational adaption and innovation.
4. The M&A integration process group seeks to address the main post-acquisition integration issues connected with the management of human resources (executives, inventors) and how these matters may affect innovation performance.
5. The articles comprising the multinational M&A research group consider M&A an effective means of entering new markets and spreading innovation thanks to knowledge transfer and spillover effects.

Our results reveal a polycentric structure of the field with few superpositions among the contributions. The multinational M&A and M&A integration process research groups present a more economics-based perspective, while the M&A innovation performance, technology-driven M&A, and M&A organizational adaption groups adopt a more strategic and organizational perspective. A partial overlap exists between the M&A innovation performance group and the multinational M&A and M&A organizational adaption groups.

The main trends concern technology-driven M&A, with a strong emphasis on organizational adaption research, somehow leaving out the M&A innovation performance literature. Instead, multinational M&A seems an evergreen area, with an apparent standby of the M&A integration process research topics. However, the combined innovation and M&A arena is still in the development phase, as observing the co-citation map's density also makes clear (Figure 3.2). In fact, the M&A integration process group, together with the M&A organizational adaption group, exhibits a high dispersion of co-citations, suggesting contents to potentially share in future investigations. We also mark particular attention to the gap in the very center of the co-citation map (Figure 3.2), which could be bridged by future contributions: for example, how the dynamics of competitive advantage (Lengnick-Hall, 1992) may be affected by technological or nontechnological M&A (Ahuja & Katila, 2001) in high-tech industries.

This work extends the existing literature in several ways. It contributes to the innovation and M&A literature by providing more accurate, thorough knowledge of their structural foundations and presenting their principal research subfields. It offers a methodological contribution of the conjoint M&A and innovation research, by applying a quantitative method based on co-citations, which allowed the definition of the intellectual structure of the field and the identification of interesting potential future trends and gaps.

Despite the above-mentioned contributions, this chapter has some limitations. The co-citation analysis provides an archival view of the research field, making it complementary to the qualitative approach (Appio, Cesaroni, & Di Minin, 2014). Moreover, this technique presents some bias regarding keyword generation and immediacy, which is strongly connected with time (Hicks, 1987, Loi et al., 2016). To overcome these issues, we selected keywords used by prior studies and chose a wide timespan, applying several thresholds (Castriotta & Di Guardo, 2016; Di Guardo & Harrigan, 2012; Small, 1977). The output of co-citation studies is highly dependent on the adoption of the database—in this case, the ISI Web of Science (Reader & Watkins, 2006). In fact, results may differ from a change in the data source; it should be recognized, however, that the ISI Web of Science database finds a high anchoring in the previous literature.

In conclusion, this work identifying the most influential literature in the examined field can be a valid support for strategy scholars and hopefully will be a source of stimulus for their future academic studies. Further research may also analyze a different time interval in order to observe changes to the intellectual structure.

ACKNOWLEDGMENTS

The work on this chapter was supported by a grant from Sardinia Regional Government (Italy) as part of the Regional Law (7/2007) "Promozione della ricerca scientifica e dell'innovazione tecnologica in Sardegna." Elona Marku gratefully acknowledges Sardinia Regional Government for the financial support of her PhD scholarship (P.O.R. Sardegna F.S.E. Operational Programme of the Autonomous Region of Sardinia, European Social Fund 2007–2013—Axis IV Human Resources, Objective l.3, Line of Activity l.3.1.).

REFERENCES

Achilladelis, B., & Antonakis, N. (2001). The dynamics of technological innovation: the case of the pharmaceutical industry. *Research Policy, 30*(4), 535–588.

Agrawal, A., & Jaffe, J. F. (2000). The post-merger performance puzzle. In C. Cooper & A. Gregory (Eds.), *Advances in mergers and acquisitions* (pp. 7–41). Greenwich, CT: JAI.

Ahuja, G., & Katila, R. (2001). Technological acquisitions and the innovation performance of acquiring firms: A longitudinal study. *Strategic Management Journal, 22*(3), 197–220.

Ailon-Souday, G., & Kunda, G. (2003). The local selves of global workers: The social construction of national identity in the face of organizational globalization. *Organization Studies, 24*(7), 1073–1096.

Allen, L., & Pantzalis, C. (1996). Valuation of the operating flexibility of multinational corporations. *Journal of International Business Studies, 27*(4), 633–653.

Appio, F. P., Cesaroni, F., & Di Minin, A. (2014). Visualizing the structure and bridges of the intellectual property management and strategy literature: A document co-citation analysis. *Scientometrics, 101*(1), 623–661.

Aybar, B., & Ficici, A. (2009). Cross-border acquisitions and firm value: An analysis of emerging-market multinationals. *Journal of International Business Studies, 40*(8), 1317–1338.

Béchard, J. P., & Grégoire, D. (2005). Entrepreneurship education research revisited: The case of higher education. *Academy of Management Learning & Education, 4*(1), 22–43.

Belderbos, R. (2003). Entry mode, organizational learning, and R&D in foreign affiliates: Evidence from Japanese firms. *Strategic management journal, 24*(3), 235–259.

Benson, D., & Ziedonis, R. H. (2009). Corporate venture capital as a window on new technologies: Implications for the performance of corporate investors when acquiring startups. *Organization Science, 20*(2), 329–351.

Blonigen, B. A., & Taylor, C. T. (2000). R&D intensity and acquisitions in high-technology industries: Evidence from the US electronic and electrical equipment industries. *Journal of Industrial Economics, 47*(1), 47–71.

Boyack, K. W., & Klavans, R. (2010). Co-citation analysis, bibliographic coupling, and direct citation: Which citation approach represents the research front most accurately? *Journal of the American Society for Information Science and Technology, 61*(12), 2389–2404.

Brau, J. C., Francis, B., & Kohers, N. (2003). The choice of IPO versus takeover: Empirical evidence. *The Journal of Business, 76*(4), 583–612.

Bresman, H. (2013). Changing routines: A process model of vicarious group learning in pharmaceutical R&D. *Academy of Management Journal, 56*(1), 35–61.

Bresman, H., Birkinshaw, J., & Nobel, R. (1999). Knowledge transfer in international acquisitions. *Journal of International Business Studies, 30*(3), 439–462.

Buckley, P. J., & Casson, M. (1996). An economic model of international joint venture strategy. *Journal of International Business Studies, 27*(5), 849–876.

Capron, L., Dussauge, P., & Mitchell, W. (1998). Resource redeployment following horizontal acquisitions in Europe and North America, 1988–1992. *Strategic Management Journal, 19*(7), 631–661.

Cartwright, S., & Schoenberg, R. (2006). Thirty years of mergers and acquisitions research: Recent advances and future opportunities. *British Journal of Management, 17*(S1), S1–S5.

Cassiman, B., Colombo, M. G., Garrone, P., & Veugelers, R. (2005). The impact of M&A on the R&D process: An empirical analysis of the role of technological- and market-relatedness. *Research Policy, 34*(2), 195–220.

Castriotta, M., & Di Guardo, M. C. (2016). Disentangling the automotive technology structure: a patent co-citation analysis. *Scientometrics, 107*(2), 819–837.

Chaudhuri, S., & Tabrizi, B. (1999). Capturing the real value in high-tech acquisitions. *Harvard Business Review, 77*(5), 123–130.

Chi, T. (1994). Trading in strategic resources: Necessary conditions, transaction cost problems, and choice of exchange structure. *Strategic Management Journal, 15*(4), 271–290.

Cloodt, M., Hagedoorn, J., & Van Kranenburg, H. (2006). Mergers and acquisitions: Their effect on the innovative performance of companies in high-tech industries. *Research Policy, 35*(5), 642–654.

Crane, D. (1972). *Invisible colleges: Diffusion of knowledge in scientific communities.* Chicago, IL: University of Chicago Press.

Crossan, M. M., & Apaydin, M. (2010). A multi-dimensional framework of organizational innovation: A systematic review of the literature. *Journal of Management Studies, 47*(6), 1154–1191.

Curran, C. S., & Leker, J. (2011). Patent indicators for monitoring convergence: Examples from NFF and ICT. *Technological Forecasting and Social Change, 78*(2), 256–273.

De Man, A. P., & Duysters, G. (2005). Collaboration and innovation: A review of the effects of mergers, acquisitions and alliances on innovation. *Technovation, 25*(12), 1377–1387.

Deng, P. (2009). Why do Chinese firms tend to acquire strategic assets in international expansion? *Journal of World Business, 44*(1), 74–84.

Di Guardo, M. C., Galvagno, M., & Cabiddu, F. (2012). Analysing the intellectual structure of e-service research. *International Journal of E-Services and Mobile Applications (IJESMA), 4*(2), 19–36.

Di Guardo, M. C., & Harrigan, K. R. (2012). Mapping research on strategic alliances and innovation: A co-citation analysis. *The Journal of Technology Transfer, 37*(6), 789–811.

Di Guardo, M. C., & Valentini, G. (2007). Explaining the effect of M&A on technological performance. In C. L. Cooper & S. Finkelstein (Eds.), *Advances in mergers and acquisitions* (pp. 107–125). Bingley, United Kingdom: Emerald.

Duysters, G., & Man, A. P. (2003). Transitory alliances: An instrument for surviving turbulent industries? *R&D Management, 33*(1), 49–58.

Eisenbeiss, S. A., van Knippenberg, D., & Boerner, S. (2008). Transformational leadership and team innovation: integrating team climate principles. *Journal of Applied Psychology, 93*(6), 1438–1446.

Ernst, H., & Vitt, J. (2000). The influence of corporate acquisitions on the behaviour of key inventors. *R&D Management, 30*(2), 105–120.

Eun, C. S., Kolodny, R., & Scheraga, C. (1996). Cross-border acquisitions and shareholder wealth: Tests of the synergy and internalization hypotheses. *Journal of Banking & Finance, 20*(9), 1559–1582.

Faleye, O., Hoitash, R., & Hoitash, U. (2011). The costs of intense board monitoring. *Journal of Financial Economics, 101*(1), 160–181.

Fichman, R. G., & Kemerer, C. F. (1999). The illusory diffusion of innovation: An examination of assimilation gaps. *Information Systems Research, 10*(3), 255–275.

Fischer, T., & Henkel, J. (2012). Patent trolls on markets for technology–An empirical analysis of NPEs' patent acquisitions. *Research Policy, 41*(9), 1519–1533.

Flatten, T. C., Engelen, A., Zahra, S. A., & Brettel, M. (2011). A measure of absorptive capacity: Scale development and validation. *European Management Journal, 29*(2), 98–116.

Frost, T. S., Birkinshaw, J. M., & Ensign, P. C. (2002). Centers of excellence in multinational corporations. *Strategic Management Journal, 23*(11), 997–1018.

Frost, T. S., & Zhou, C. (2005). R&D co-practice and 'reverse' knowledge integration in multinational firms. *Journal of International Business Studies, 36*(6), 676–687.

Gambardella, A., & Torrisi, S. (1998). Does technological convergence imply convergence in markets? Evidence from the electronics industry. *Research Policy, 27*(5), 445–463.

Garfield, E. (1979). Is citation analysis a legitimate evaluation tool? *Scientometrics, 1*(4), 359–375.

Goodman, L. E., & Dion, P. A. (2001). The determinants of commitment in the distributor–manufacturer relationship. *Industrial Marketing Management, 30*(3), 287–300.

Gopalakrishnan, S., & Damanpour, F. (1997). A review of innovation research in economics, sociology and technology management. *Omega, 25*(1), 15–28.

Graebner, M. E. (2004). Momentum and serendipity: How acquired leaders create value in the integration of technology firms. *Strategic Management Journal, 25*(8–9), 751–777.

Graves, S. B., & Langowitz, N. S. (1993). Innovative productivity and returns to scale in the pharmaceutical industry. *Strategic Management Journal, 14*(8), 593–605.

Groysberg, B., Lee, L. E., & Nanda, A. (2008). Can they take it with them? The portability of star knowledge workers' performance. *Management Science, 54*(7), 1213–1230.

Gupta, A. K., Smith, K. G., & Shalley, C. E. (2006). The interplay between exploration and exploitation. *Academy of Management Journal, 49*(4), 693–706.

Hagedoorn, J., & Cloodt, M. (2003). Measuring innovative performance: Is there an advantage in using multiple indicators? *Research Policy, 32*(8), 1365–1379.

Hagedoorn, J., & Wang, N. (2012). Is there complementarity or substitutability between internal and external R&D strategies? *Research Policy, 41*(6), 1072–1083.

Håkanson, L., & Nobel, R. (1993). Foreign research and development in Swedish multinationals. *Research Policy, 22*(5), 373–396.

Hambrick, D. C., & Cannella, A. A. (1993). Relative standing: A framework for understanding departures of acquired executives. *Academy of Management journal, 36*(4), 733–762.

Han, K., Oh, W., Im, K. S., Chang, R. M., Oh, H., & Pinsonneault, A. (2012). Value cocreation and wealth spillover in open innovation alliances. *MIS Quarterly, 36*(1), 291–325.

Haunschild, P. R., & Miner, A. S. (1997). Modes of interorganizational imitation: The effects of outcome salience and uncertainty. *Administrative Science Quarterly, 42*(3), 472–500.

Hayward, M. L. (2002). When do firms learn from their acquisition experience? Evidence from 1990–1995. *Strategic Management Journal, 23*(1), 21–39.

Healy, P. M., Palepu, K. G., & Ruback, R. S. (1992). Does corporate performance improve after mergers? *Journal of Financial Economics, 31*(2), 135–175.

Hicks, D. (1987). Limitations of co-citation analysis as a tool for science policy. *Social Studies of Science, 17*(2), 295–316.

Higgins, M. J., & Rodriguez, D. (2006). The outsourcing of R&D through acquisitions in the pharmaceutical industry. *Journal of Financial Economics, 80*(2), 351–383.

Hitt, M. A., Hoskisson, R. E., & Ireland, R. D. (1994). A mid-range theory of the interactive effects of international and product diversification on innovation and performance. *Journal of Management, 20*(2), 297–326.

Hitt, M. A., Hoskisson, R. E., Johnson, R. A., & Moesel, D. D. (1996). The market for corporate control and firm innovation. *Academy of Management Journal, 39*(5), 1084–1119.

Hoskisson, R. E., Hitt, M. A., & Hill, C. W. (1993). Managerial incentives and investment in R&D in large multiproduct firms. *Organization Science, 4*(2), 325–341.

Hoskisson, R. E., Hitt, M. A., Johnson, R. A., & Moesel, D. D. (1993). Construct validity of an objective (entropy) categorical measure of diversification strategy. *Strategic Management Journal, 14*(3), 215–235.

Karim, S. (2006). Modularity in organizational structure: The reconfiguration of internally developed and acquired business units. *Strategic Management Journal, 27*(9), 799–823.

Karim, S., & Mitchell, W. (2000). Path-dependent and path-breaking change: Reconfiguring business resources following acquisitions in the U.S. medical sector, 1978–1995. *Strategic Management Journal, 21*(10/11), 1061–1081.

Kelm, K. M., Narayanan, V. K., & Pinches, G. E. (1995). Shareholder value creation during R&D innovation and commercialization stages. *Academy of Management Journal, 38*(3), 770–786.

King, D. R., Dalton, D. R., Daily, C. M., & Covin, J. G. (2004). Meta-analyses of post-acquisition performance: Indications of unidentified moderators. *Strategic Management Journal, 25*(2), 187–200.

Kitchener, M. (2002). Mobilizing the logic of managerialism in professional fields: The case of academic health centre mergers. *Organization Studies, 23*(3), 391–420.

Lambe, C. J., & Spekman, R. E. (1997). Alliances, external technology acquisition, and discontinuous technological change. *Journal of Product Innovation Management, 14*(2), 102–116.

Lengnick-Hall, C. A. (1992). Innovation and competitive advantage: What we know and what we need to learn. *Journal of Management, 18*(2), 399–429.

Li, Y. R. (2009). The technological roadmap of Cisco's business ecosystem. *Technovation, 29*(5), 379–386.

Loi, M., Castriotta, M., & Di Guardo, M. C. (2016). The theoretical foundations of entrepreneurship education: How co-citations are shaping the field. *International Small Business Journal, 34*(7), 948–971.

Long, W. F., & Ravenscraft, D. J. (1993). LBOs, debt and R&D intensity. *Strategic Management Journal, 14*, 119–135.

Makadok, R. (2001). Toward a synthesis of the resource-based and dynamic-capability views of rent creation. *Strategic Management Journal, 22*(5), 387–401.

Makri, M., Hitt, M. A., & Lane, P. J. (2010). Complementary technologies, knowledge relatedness, and invention outcomes in high technology mergers and acquisitions. *Strategic Management Journal, 31*(6), 602–628.

March, J. G. (1991). Exploration and exploitation in organizational learning. *Organization Science, 2*(1), 71–87.

Marsh, S. J., & Stock, G. N. (2003). Building dynamic capabilities in new product development through intertemporal integration. *Journal of Product Innovation Management, 20*(2), 136–148.

Marsh, S. J., & Stock, G. N. (2006). Creating dynamic capability: The role of intertemporal integration, knowledge retention, and interpretation. *Journal of Product Innovation Management, 23*(5), 422–436.

McCain, K. W. (1990). Mapping authors in intellectual space: A technical overview. *Journal of the American Society for Information Science (1986–1998), 41*(6), 433–443.

Morosini, P., Shane, S., & Singh, H. (1998). National cultural distance and cross-border acquisition performance. *Journal of International Business Studies, 29*(1), 137–158.

Nemanich, L. A., & Vera, D. (2009). Transformational leadership and ambidexterity in the context of an acquisition. *The Leadership Quarterly, 20*(1), 19–33.

Nerur, S. P., Rasheed, A. A., & Natarajan, V. (2008). The intellectual structure of the strategic management field: An author co-citation analysis. *Strategic Management Journal, 29*(3), 319–336.

Newman, M. E. J. & Girvan, M. 2004 Finding and evaluating community structure in networks. *Physical Review E 69*(2):026113. doi:10.1103/PhysRevE.69.026113

Ofek, E., & Sarvary, M. (2001). Leveraging the customer base: Creating competitive advantage through knowledge management. *Management Science, 47*(11), 1441–1456.

Officer, M. S., Poulsen, A. B., & Stegemoller, M. (2009). Target-firm information asymmetry and acquirer returns. *Review of Finance, 13*(3), 467–493.

O'Neill, H. M., Pouder, R. W., & Buchholtz, A. K. (1998). Patterns in the diffusion of strategies across organizations: Insights from the innovation diffusion literature. *Academy of Management Review, 23*(1), 98–114.

O'Reilly, C., Tushman, M., & Harreld, J. B. (2009). Organizational ambidexterity: IBM and emerging business opportunities. *California Management Review, 51,* 75–99

Paruchuri, S., Nerkar, A., & Hambrick, D. C. (2006). Acquisition integration and productivity losses in the technical core: Disruption of inventors in acquired companies. *Organization Science, 17*(5), 545–562.

Pennings, J. M., Barkema, H., & Douma, S. (1994). Organizational learning and diversification. *Academy of Management journal, 37*(3), 608–640.

Ponzi, L. J. (2002). The intellectual structure and interdisciplinary breadth of knowledge management: A bibliometric study of its early stage of development. *Scientometrics, 55*(2), 259–272.

Porrini, P. (2004). Can a previous alliance between an acquirer and a target affect acquisition performance? *Journal of management, 30*(4), 545–562.

Prabhu, J. C., Chandy, R. K., & Ellis, M. E. (2005). The impact of acquisitions on innovation: Poison pill, placebo, or tonic? *Journal of Marketing, 69*(1), 114–130.

Puranam, P., Singh, H., & Zollo, M. (2006). Organizing for innovation: Managing the coordination-autonomy dilemma in technology acquisitions. *Academy of Management Journal, 49*(2), 263–280.

Puranam, P., & Srikanth, K. (2007). What they know vs. what they do: How acquirers leverage technology acquisitions. *Strategic Management Journal, 28*(8), 805–825.

Raisch, S., Birkinshaw, J., Probst, G., & Tushman, M. L. (2009). Organizational ambidexterity: Balancing exploitation and exploration for sustained performance. *Organization Science, 20*(4), 685–695.

Ramos-Rodríguez, A. R., & Ruíz-Navarro, J. (2004). Changes in the intellectual structure of strategic management research: A bibliometric study of the *Strategic Management Journal,* 1980–2000. *Strategic Management Journal, 25*(10), 981–1004.

Ranft, A. L., & Lord, M. D. (2002). Acquiring new technologies and capabilities: A grounded model of acquisition implementation. *Organization Science, 13*(4), 420–441.

Reader, D., & Watkins, D. (2006). The social and collaborative nature of entrepreneurship scholarship: A co-citation and perceptual analysis. *Entrepreneurship theory and practice, 30*(3), 417–441.

Reitzig, M., & Puranam, P. (2009). Value appropriation as an organizational capability: The case of IP protection through patents. *Strategic Management Journal, 30*(7), 765–789.

Rothaermel, F. T., & Thursby, M. (2007). The nanotech versus the biotech revolution: Sources of productivity in incumbent firm research. *Research Policy, 36*(6), 832–849.

Salvato, C., Chirico, F., & Sharma, P. (2010). A farewell to the business: Championing exit and continuity in entrepreneurial family firms. *Entrepreneurship and Regional Development, 22*(3–4), 321–348.

Santos, D. F. L., Basso, L. F. C., Kimura, H., & Kayo, E. K. (2014). Innovation efforts and performances of Brazilian firms. *Journal of Business Research, 67*(4), 527–535.

Seru, A. (2014). Firm boundaries matter: Evidence from conglomerates and R&D activity. *Journal of Financial Economics, 111*(2), 381–405.

Sher, P. J., & Yang, P. Y. (2005). The effects of innovative capabilities and R&D clustering on firm performance: the evidence of Taiwan's semiconductor industry. *Technovation, 25*(1), 33–43.

Sitkin, S. B. (1992). Learning through failure-The strategy of small losses. *Research in Organizational Behavior, 14*, 231–266.

Small, H. (1973). Co-citation in the scientific literature: A new measure of the relationship between two documents. *Journal of the American Society for information Science, 24*(4), 265–269.

Small, H. G. (1977). A co-citation model of a scientific specialty: A longitudinal study of collagen research. *Social Studies of Science, 7*(2), 139–166.

Small, H. (2003). Paradigms, citations, and maps of science: A personal history. *Journal of the American Society for information Science and Technology, 54*(5), 394–399.

Small, H. (2010). Maps of science as interdisciplinary discourse: Co-citation contexts and the role of analogy. *Scientometrics, 83*(3), 835–849.

Stern, D. I. (2014). High-ranked social science journal articles can be identified from early citation information. *PLoS ONE, 9*(11), e112520

Stuart, T. E., & Sorenson, O. (2003). Liquidity events and the geographic distribution of entrepreneurial activity. *Administrative Science Quarterly, 48*(2), 175–201.

Tsang, E. W., & Yip, P. S. (2007). Economic distance and the survival of foreign direct investments. *Academy of Management Journal, 50*(5), 1156–1168.

Tushman, M. L., & O'Reilly III, C. A. (1996). Managing evolutionary and revolutionary change. *California Management Review, 38*(4), 8–28.

Vaara, E., & Monin, P. (2010). A recursive perspective on discursive legitimation and organizational action in mergers and acquisitions. *Organization Science, 21*(1), 3–22.

Valentini, G., & Di Guardo, M. C. (2012). M&A and the profile of inventive activity. *Strategic Organization, 10*(4), 384–405.

van de Vrande, V., Lemmens, C., & Vanhaverbeke, W. (2006). Choosing governance modes for external technology sourcing. *R&D Management, 36*(3), 347–363.

van de Vrande, V., Vanhaverbeke, W., & Duysters, G. (2009). External technology sourcing: The effect of uncertainty on governance mode choice. *Journal of business venturing, 24*(1), 62–80.

van Eck, N. J., & Waltman, L. (2007). Bibliometric mapping of the computational intelligence field. *International Journal of Uncertainty, Fuzziness and Knowledge-Based Systems, 15*(5), 625–645.

van Eck, N., & Waltman, L. (2009). Software survey: VOSviewer, a computer program for bibliometric mapping. *Scientometrics, 84*(2), 523–538.

van Eck, N. J., Waltman, L., Dekker, R., & van den Berg, J. (2010). A comparison of two techniques for bibliometric mapping: Multidimensional scaling and VOS. *Journal of the American Society for Information Science and Technology, 61*(12), 2405–2416.

van Eck, N. J., Waltman, L., Den Berg, J., & Kaymak, U. (2006). Visualizing the computational intelligence field [Application Notes]. *Computational Intelligence Magazine, IEEE, 1*(4), 6–10.

Vanhaverbeke, W., Duysters, G., & Noorderhaven, N. (2002). External technology sourcing through alliances or acquisitions: An analysis of the application-specific integrated circuits industry. *Organization Science, 13*(6), 714–733.

Vassolo, R. S., Anand, J., & Folta, T. B. (2004). Non-additivity in portfolios of exploration activities: A real options-based analysis of equity alliances in biotechnology. *Strategic Management Journal, 25*(11), 1045–1061.

Waaijer, C. J., van Bochove, C. A., & van Eck, N. J. (2010). Journal editorials give indication of driving science issues. *Nature, 463*(7278), 157–157.

Wang, L., & Zajac, E. J. (2007). Alliance or acquisition? A dyadic perspective on interfirm resource combinations. *Strategic Management Journal, 28*(13), 1291–1317.

White, H. D., & McCain, K. W. (1998). Visualizing a discipline: An author co-citation analysis of information science, 1972–1995. *Journal of the American Society for Information Science, 49*(4), 327–355.

Wolfe, R. A. (1994). Organizational innovation: Review, critique and suggested research directions. *Journal of Management Studies, 31*(3), 405–431.

Zahra, S. A. (1995). Corporate entrepreneurship and financial performance: The case of management leveraged buyouts. *Journal of Business Venturing, 10*(3), 225–247.

Zahra, S. A., & Hayton, J. C. (2008). The effect of international venturing on firm performance: The moderating influence of absorptive capacity. *Journal of Business Venturing, 23*(2), 195–220.

Zollo, M., & Reuer, J. J. (2010). Experience spillovers across corporate development activities. *Organization Science, 21*(6), 1195–1212.

CHAPTER 4

TECHNOLOGICAL OPPORTUNITIES AND THEIR REJECTION

A Process Perspective on Organizational Lock-in

Wasko Rothmann and Matthias Wenzel
European University Viadrina, Frankfurt (Oder)

Heinz-Theo Wagner
German Graduate School of Management and Law, Heilbronn

ABSTRACT

In this chapter, we add to the contemporary debate on path dependence in the management literature by examining how the scope of action evolves in lock-ins. For this purpose, we draw on qualitative data of the German newspaper industry from 1995 to date. Our study indicates that the scope of action in lock-ins continuously changes as new technological opportunities emerge and disappear. Our results also suggest that path-dependent firms have windows of opportunity available that remain unexploited due to cognitive and

Technological Innovation Networks, pages 75–97

normative barriers. Furthermore, our empirical evidence indicates that although windows of opportunity may temporarily close, the scope of action reopens when further technological opportunities emerge. Our study contributes to a better understanding of lock-ins as an underresearched phase of path dependence, provides a more comprehensive process understanding for the inability of path-dependent firms to exploit technological opportunities, and offers practical implications on how path-dependent firms may leave their path.

Debates on path dependence have recently gained momentum in the management literature (e.g., Dobusch & Kapeller, 2013; Dobusch & Schüßler, 2013; Koch, 2008, 2011; Rothmann & Koch, 2014; Schmidt & Braun, 2015; Schreyögg & Sydow, 2011; Sydow, Schreyögg, & Koch, 2009; Sydow, Windeler, Müller-Seitz, & Lange, 2012; Sydow, Windeler, Müller-Seitz, & Lange, 2012; Vergne & Durand, 2011; Wenzel, 2015). Whereas scholars in other disciplines such as economics, sociology, and political science have developed a consolidated understanding of path dependence (Arthur, 1989; David, 1985; Mahoney, 2000; Pierson, 2000; Shapiro & Varian, 1999), management scholars have long referred only loosely to this concept (Vergne & Durand, 2011). With its recent theoretical specification in the management context (Sydow et al., 2009), management scholars have come to understand path dependence as a process in which managerial scope of action is reduced over time because of the enactment of positive feedback (Dobusch & Kapeller, 2013). As a result of this process, affected firms may become "locked-in"; that is, they have no remaining scope of action and are unable to exploit emerging technological opportunities (Sydow et al., 2009; Vergne & Durand, 2011). With that, path dependence theory provides a nascent theoretical perspective on a contemporary phenomenon that has befallen entire industries, such as the music industry (Kunow, Gersch, & Koch, 2013), the newspaper industry (Rothmann & Koch, 2014), the book industry (Schreyögg, Sydow, & Holtmann, 2011), and the funeral industry (Wenzel, 2015).

Whereas other theoretical approaches to inertia (e.g., Hannan & Freeman, 1977; Staw, 1981; Stinchcombe, 1965) do not fully explicate how organizational rigidity emerges, path dependence—with self-reinforcing mechanisms "at the heart" (Sydow et al., 2009, p. 698)—provides compelling answers to this question (Koch, 2008). Therefore, path dependence literature has predominantly focused on how paths emerge (Dobusch & Schüßler, 2013). Accordingly, the extant literature on path dependence pertinently sheds light on the nonlinear dynamics that lead to the establishment of a dominant pattern of strategic choice (Schreyögg & Sydow, 2011). However, only few studies focus especially on the lock-in stage of path dependence in which firms are unable to exploit emerging technological opportunities (Rothmann & Koch, 2014). These studies holistically refer to three reasons for a constrained scope of action that inhibits firms from exploiting emerging technological opportunities: resource-based, cognitive, and normative (Koch, 2008, 2011; Sydow et al., 2009). Although this classification of reasons provides valuable insights into firms' inability to exploit emerging technological opportunities, it does not fully account for the processual nature of path dependence (Schreyögg & Sydow, 2011); that is, whereas this classification reveals specific reasons for sticking to an established choice pattern in specific points in time, the idea of path dependence suggests that the reasons for rejecting the exploitation of emerging technological opportunities may change over time, leading to alternating opportunities to leave the established path (Sydow, Windeler, Müller-Seitz, & Lange, 2012). Therefore, we aim to gain a process understanding of the lock-in stage of path dependence by addressing the following research question: *How does the scope of action evolve in lock-ins?*

We explore this phenomenon in a case study of the German high-quality daily newspaper publishing industry. This industry can be considered paradigmatic for news publishers all over the world whose traditional, paper-based business models have been put into question by online news offerings (Gilbert, 2005). Furthermore, the newspaper industry has still been unable to exploit emerging technological opportunities that have come along with the digitization of news products (Oestreicher-Singer & Zalmanson, 2013), such as the possibility of producing and marketing journalistic content and ads separately, developing a willingness to pay for digital news early on and establishing payment technologies that ease the process of paying for digital news, and developing differentiated news products for mobile and tablet systems. Our case analysis indicates that the reasons such technological opportunities are left unexploited continuously alternate between cognitive, normative, and resource-based forms. Furthermore, our findings suggest that path-dependent firms do not necessarily suffer from a lack of technological opportunities; instead, when technological opportunities emerge, path-dependent firms may have a broader scope of action available and dispose of "windows of opportunity" (Tyre & Orlikowski, 1994) in which they could exploit emerging technological opportunities. However, our empirical evidence indicates that these technological opportunities remain unexploited because path-dependent firms reject them for cognitive or normative reasons until resource-based reasons prevail and windows of opportunity close. Nevertheless, our findings also suggest that windows of opportunity close only temporarily and reopen when new technological opportunities emerge.

Our findings contribute both theoretical and practical insights to the contemporary debate on path dependence in the management literature. First, our study complements the extant literature on path dependence that has mostly illuminated the emergence of paths (Dobusch & Schüßler, 2013) by shedding further light on the lock-in stage of path dependence that has remained underrepresented so far (Rothmann & Koch, 2014). In particular, we examine the evolution of the scope of action in the lock-in stage and find an alternation of the range of available options that comes along with the emergence of technological opportunities. Second, our chapter offers a process understanding of the lock-in stage of path dependence. Thus, our study extends beyond static ascriptions of reasons for firms' inability to exploit technological opportunities that prevail in the literature on path dependence (e.g., Sydow et al., 2009) by providing a process perspective that provides a more comprehensive answer to the question of why path-dependent firms are unable to exploit emerging technological opportunities. In particular, our study suggests that the scope of action continuously alternates between forms of lock-in that can be overcome and forms that are hardly escapable. Third, whereas the extant literature on path dependence draws a pessimistic picture with regard to overcoming path dependence in lock-ins (Rothmann & Koch, 2014), our process view on the lock-in stage outlines that path-dependent firms have windows of opportunity available to escape lock-in. Thus, in practical terms, our analysis indicates that path-dependent firms, such as news publishers, may receive further opportunities to escape their path despite a number of missed technological opportunities. With the help of external interventions (e.g., Sydow et al., 2009), these firms may overcome cognitive and normative barriers to become sensitive to the emergence of further technological opportunities.

THEORETICAL BACKGROUND: PATH DEPENDENCE AND FORMS OF LOCK-IN

Path dependence is, first and foremost, a process (Schreyögg & Sydow, 2011). In an organizational context, it explains how the scope of action to exploit technological opportunities may narrow over time until it is tightly constricted in the resulting lock-in. Strategic choice thereby converges to an ultimately predominant pattern over three phases (Sydow et al., 2009).

In the first stage, the preformation phase, firms have a wide corridor of options available that are only slightly limited by historical imprints. A seemingly small event (Arthur, 1989) may eventually induce the second stage, the formation phase, in which the range of variety gradually becomes constricted. This process is essentially driven by self-reinforcing mechanisms, such as complementarity effects, learning effects, coordination effects, adaptive expectations, network effects, and economies of scale (Dobusch & Schüßler, 2013). These mechanisms gradually drive out strategic alternatives due to positive feedback on the selection of certain strategic alternatives, such that they increase the likelihood of the emergence of an ultimately predominant pattern of strategic choice. These strategic decisions are generally still reversible but more difficult to revoke toward the end of this phase (Miller, 1993). Therefore, the scope of action is constricted even further. If self-reinforcing mechanisms drive out most strategic alternatives, firms enter the third stage, the lock-in phase. In this case, firms are bound to the predominant strategic pattern that has emerged and only have a narrow range of variety left.

The resulting lock-in is not critical to firm activity in all cases. Instead, the self-reinforcing dynamics of path dependence lead firms to a strategic fit (Zajac, Kraatz, & Bresser, 2000) of the reproduced pattern with specific environmental conditions. Thus, the strategic pattern may be highly efficient and may be reproduced successfully over a long period of time (Rothmann & Koch, 2014). However, lock-ins constitute a major challenge for firms when the environment changes (see also Wohlgemuth & Wenzel, 2016). As the scope of action is sacrificed for increased efficiency of the strategic pattern under given environmental conditions, path-dependent firms hardly perceive any scope of action to adjust their strategic pattern (March, 1991). This rigidity prevents firms from exploiting new technological opportunities and may lead to organizational decline (Leonard-Barton, 1992).

In principle, environmental changes broaden the corridor of choice alternatives that are available to firms by jeopardizing previous successes (Arthur, 1994). Correspondingly, emerging technological opportunities constitute large strategic potentials (Drnevich & Croson, 2013) even for firms that are locked in. However, the exploitation of new technological opportunities seems to be particularly challenging in lock-ins (Pentland, Feldman, Becker, & Liu, 2012). Path-dependent firms therefore tend to reproduce the stable strategic pattern for a long period of time even after environmental changes render the choice pattern inefficient (Rothmann & Koch, 2014; Wenzel, 2015). Path dependence theory provides specific reasons why firms do not opt for attractive strategic alternatives. In particular, the concept differentiates between resource-based, cognitive, and normative forms of lock-in (Koch, 2008, 2011; Sydow et al., 2009).

In the case of a *resource-based lock-in*, path-dependent firms dispose of a narrowed scope of action because their entire resource base is tightly aligned with specific environmental conditions. This specificity results from increasing returns on investing in and exploiting the existing resource base to reproduce the established strategic pattern (March, 1991). The required accumulation of new resources renders new strategic options unattractive, such that firms in resource-based lock-ins virtually have no technological opportunities available (Arthur, 1994).

Firms in *cognitive lock-ins* may have attractive strategic alternatives available, but they do not perceive these as such. Along with the increasing success of the emerged strategic pattern, the reconstructed success factors get deeply inscribed into the cognitive structures and are increasingly considered to be unchangeable (Tripsas & Gavetti, 2000) or kept in place by rigidifying mechanisms (Wenzel, 2015). Therefore, affected firms do not recognize viable strategic options as part of their scope of action, even though they may have the resources required to exploit these technological opportunities.

If firms recognize viable strategic options but still reject them, they are trapped in a *normative lock-in*. In this case, affected firms gradually identify with the business activities determined by the increasingly successful strategic pattern (Tripsas & Gavetti, 2000). These firms therefore opt for those strategic alternatives that do not threaten their identity, whereas the continuation of the strategic pattern is most likely (Gilbert, 2005). However, particularly in normative lock-ins, self-reinforcing mechanisms may have driving effects even beyond the formation phase. For instance, adaptive expectations may lead to the rejection of alternatives that are part of their scope of action if adverse reactions by consumers and/or competitors are expected (Danneels, 2003).

RESEARCH METHODOLOGY

Research Design and Case Selection

To explore how the scope of action evolves in lock-ins, we conducted a retrospective qualitative case study (Yin, 2014) whose design resembles other insightful studies that analyze the inability of firms to conduct strategic change in times of technological shifts (e.g., Burgelman, 1991; Tripsas & Gavetti, 2000). In particular, we analyzed the case of the German newspaper industry in light of path dependence and the emergence of technological opportunities. Starting in 2001, the newspaper industry around the world experienced a major crisis, driven by declining advertising markets due to the rupture of the "dot-com bubble" (Gilbert, 2005). The German newspaper market—formerly one of the world's largest and most diverse newspaper markets (BDZV, 2005)—came under significant pressure, such that numerous newspapers were discontinued and the remaining news publishers drastically reduced content production (BDZV, 2003). Despite the insightful discourse on the characteristics of viable business models for the commercialization of digital news (e.g., Lopes & Galletta, 2006; Oestreicher-Singer & Zalmanson, 2013), none of the traditional big players in the German newspaper industry has found a sustainable way to transform the newspaper business model (Rothmann & Koch, 2014). Due to the severity of market disruptions that caused urgent needs to transform the traditional business model and the German news publishers' inability to do so because of path-dependent dynamics (Koch, 2008, 2011; Rothmann & Koch, 2014), the German newspaper industry served as a revelatory case in which we could observe the focal phenomenon particularly well (see Yin, 2014).

Our research question innately focuses on the emergence of technological opportunities that firms in lock-ins may exploit or reject. All players in the newspaper industry had the same opportunities and, although they partially debated them, they decided not to exploit them. Due to the convergence of firms' strategic activities in this industry (Koch, 2008), we analyzed the strategic activities of individual players jointly as a holistic single case (Langley, 1999; Yin, 2014) with the industry's lock-in as the unit of analysis. In particular, we were interested in the industry discourse (Phillips & Hardy, 2002) on the scope of action in light of emerging technological opportunities in order to stipulate which strategic options are (or are not) debated and which opportunities are (or are not) exploited over time (Koch, 2011). For our analysis, we focused on the so-called "daily quality press," which is a paradigmatic "lighthouse" (Blum, Bonfadelli, Imhof, & Jarren, 2011) of the industry with an aggregated paid circulation of 1.3 million units in 2000 (IVW, 2017) and well-suited for investigation for mainly two reasons: (a) production costs are higher in this segment than in the overall industry (BDZV, 2002) and, therefore, the need for change rapidly became more urgent during the industry crisis, and (b) the newspapers in this segment were/are regarded as journalistic flagships (Blum et al., 2011), such that the industry discourse was mainly focused on their crisis. In the next section, we describe the context of our case in more detail.

Data Collection and Analysis

In order to facilitate data triangulation, we gathered data from multiple sources (Yin, 2014). First, we browsed specialist publications—mainly six trade newspapers and magazines, such as *W&V* and *Horizont*—that observe and report developments in the newspaper industry in detail. We read all articles that were found and saved 1,986 relevant ones for further analysis. We then conducted 27 retrospective semi-structured interviews in all (five) nationwide published German daily quality newspapers between 2010 and 2013. Most of the interviewees were key decision-makers in the strategy process of these firms (6 CEOs, 3 C-levels, 11 senior executives, 5 middle managers, and 2 industry experts) and reflected both the strategic actions taken and the shared rationale behind these actions (Lincoln & Guba, 1985). The interviews focused on industry dynamics since 1995 and lasted 1.25 hours on average. All but one interview were audio-recorded and transcribed for further analysis. One interviewee refused audio-recording. We recorded this interview in writing, and the interview subsequently confirmed the protocol. In an iterative loop, we conducted five additional interviews following our first data analysis to validate our findings.

The analysis of collected data proceeded in four stages. First, we leveraged the strength of the retrospective research design—the possibility of gathering rich data even for lengthy transformation processes that are difficult to observe in real time (Sydow, Windeler, Schubert, & Möllering, 2012a)—to construct a narrative story of the chronological development (Langley, 1999) of the news publishers' lock-in that has evolved for more than a decade. This "thick description" (Geertz, 1973) provided us with a rich understanding of this development and enabled us to analyze the distinct events based on path dependence theory.

Second, we decomposed this chronology into analytic periods (Van de Ven, 1992), which distinguished the dominant forms of lock-in in which the news publishers were trapped in different periods. This step facilitated a more fine-grained analysis of each phase and allowed us to better characterize the scope of action.

Third, we coded the entire material across these phases to analyze the evolution of the scope of action using the qualitative data analysis software *Atlas.ti*, v6. For this purpose, we developed and used 135 codes (available upon request), which we derived deductively based on proposed frameworks to analyze paths (Koch, 2011; Sydow et al., 2012a; see also Wenzel, 2015; Wenzel, Senf, & Koch, 2016) and which partially emerged inductively from the data (Corbin & Strauss, 1990). With these categories, we coded 6,640 instances.

Fourth, when we obtained preliminary research findings, we conducted five follow-up interviews with key informants in which we presented and discussed our findings. In light of the weakness of the retrospective research design to bear risks of ex-post rationalization (Yin, 2014), we could validate and ensure the trustworthiness of our research findings (Lincoln & Guba, 1985).

The Initial Situation: The Newspaper Lock-in and Its Crisis

News publishers have traditionally relied on two markets to generate revenue: that is, the readership market and the advertising market. Building on the high complementarity of both markets in the newspaper context, several positive feedback mechanisms created the so-called *advertising-circulation spiral* (Hoskins, 2004). Concisely, the spiral enabled publishing houses to subsidize their product—which has always evoked large production costs (BDZV, 2002)—by reducing the purchase price for readers or by improving the quality of the newspaper product. Each of those actions would most likely increase the readership's demand for the product and, therefore, enhance product circulation. A higher circulation would simultaneously lead to increased demand in the advertising market, resulting in higher advertising revenue, which again can be used to produce better or more content for a lower price in the readership market. Thus, this spiral drove the self-reinforcing dynamics of path dependence in this industry (Rothmann & Koch, 2014).

The advertising-circulation spiral, which operated efficiently until 2001, enabled the news publishers to reduce the newspapers' price for the readers by up to 70 percent: Without advertising revenue, a printed newspaper would have cost about three times as much (Dovifat, 1967). The attractiveness of subsidizing newspaper production with advertising revenue resulted in a lock-in in which the news publishers focused solely on the generation of revenue in the advertising market (Rothmann & Koch, 2014).

The news publishers' traditional business model of focusing on advertising revenue to offer a subsidized product to readers began to fail at the end of the 20th century, when news publishers experienced increasing difficulties attracting a younger readership as subscribers of their newspapers. Up to 2001, the news publishers could compensate the resulting decline in revenue in the readership market by generating revenue in a strongly growing advertising market. However, when the advertising market declined due to the rupture of the dot-com bubble in 2001, the business model faced a deep crisis: News publishers dramatically lost both readership and advertising revenue (BDZV, 2002; ZAW, 2004).

FINDINGS: MULTIPLE WINDOWS OF OPPORTUNITY FOR REINVENTING THE NEWSPAPER BUSINESS MODEL

In this section, we present the findings of our study. We identified five periods (1995–2000, 2001–2007, 2008–2009, 2010–2011, since 2012), three of which are characterized by different windows of opportunity that all remained unexploited. For every period, we describe the changes of the scope of action in light of emerging technological opportunities and explain why this emerging scope of action remained unutilized by referring to prevailing forms of lock-in. Table 4.1 provides an overview of the findings for all investigated time periods.

1995–2000: The First Window of Opportunity: Separating Content and Ads in the Digital Sphere

When the internet evolved to become a mass medium in the 1990s, news publishers had a major opportunity to reinvent their business model by

TABLE 4.1 Overview of the News Publishers' Lock-in Situations

Description of news publishers' lock-in	Representative supporting data
Time period 1: 1995–2000, cognitive lock-in • News publishers began developing digital businesses in the second half of the 1990s. • Compared with the printed product, online products were seen as unattractive both in advertising and consumer environments. • Furthermore, publishers could not imagine that some of their core businesses in the advertising market could also exist without producing content. • When firms like the "Scout Group" started online classified ad portals, news publishers did not believe in their success, perceiving themselves as offering superior products in both the advertising market and the readership market. • However, news publishers felt the need to be present on the web with their newspaper brands. • Free online content was regarded as a suitable self-advertisement for the printed products. *Publishing paid digital news without the need for major advertising revenue was not recognized as a potential strategic alternative.*	Manager 6: "Of course, you can go to a news page but news consumption is quite different there. You click on and on but you won't get the deep overview like in the printed product." Manager 2: "[Y]ou could not imagine that there is someone who will be able to corrupt your business model. And who was Scout when being founded—a few people and a bit [of] money from Beisheim, that's it." Manager 16: "There was no business model in the 1990s but it was obvious that a journalistic brand has to be present in the online segment."
Time period 2: 2001–2007, normative lock-in • When the dot-com bubble burst in 2001, news publishers were confronted with a massive loss in advertising revenue. • After 2001, the online advertising market grew again, whereas revenue in the print advertising market remained low. • News publishers began attempting to utilize digitization for recovering their advertising business. • They developed free online content strategies to maximize the coverage of online news brands with the aim to regain attractiveness in the advertising market. *Publishing paid digital news was perceived as disadvantageous for trying to recover the news brands' advertising markets and was therefore not seen as a suitable strategic alternative.*	Manager 12: "The newspaper is in a structural declining process.... It is very important to gain [digital advertising] market share to compensate this effect." Manager 28 (trade press interview): "The future of digital media will be about catching up with the world's Googles and eBays." Manager 22: "I think, if a publisher had invented eBay, we would not have today's problems."

(continued)

TABLE 4.1 Overview of the News Publishers' Lock-in Situations (continued)

Description of news publishers' lock-in	Representative supporting data
Time period 3: 2008–2009, resource-based lock-in • A second major advertising crisis in 2008, deeply linked to the worldwide financial crisis, made clear that news publishers would most likely be unable to earn enough advertising revenue in the digital sphere to refinance the efforts in producing high-quality content. • Consequently, an intensive paid-content debate started. • News publishers exposed themselves to a comprehensive and deeply inscribed free content culture of news consumption. • News publishers noticed that it was probably a mistake to publish online news for free but that the effects were irreversible in the short term. *Publishing paid digital news was hardly realizable in the short or even medium term because of the established online free news consumption culture. Publishers had no suitable resources (i.e., content or technology) to overcome or change these cultural restrictions. As of 2008, paid online news content was not a viable strategic alternative.*	Manager 19: "Everyone says so: In a certain sense, free content is a birth defect of the Internet. You can't correct a birth defect." Manager 28 (trade press interview): "The free content culture is a dramatic mistake of early decisions." Manager 21: "You can calculate the potential loss in coverage [caused by an online a pay wall]: We have examples for minus 80 percent."
Time period 4: 2010–2011, normative lock-in • Tablet systems and smartphones dissolved many of the limiting technological restrictions that news publishers were exposed to. • However, at the same time, the advertising market recovered again and the news publishers recognized the opportunity to expand their brands by adding app products to their digital portfolio. • The tablet PC was perceived as a very attractive product for advertisers. • However, paid news apps and (nearly the same) free online news content co-existed. • Even for tablet apps, free content models were intensively debated as a further option to generate revenue in the advertising market. *Paid apps were not evaluated as a strategic alternative to (free) online news portals. Due to the regaining advertising market, focusing on the advertising potentials of new technologies for news publishers was (again) perceived as the superior choice compared with attempts to change the free content culture.*	Manager 12: "There are many media businesses that work fantastic with free content, e.g., TV [in Germany], radio, the whole Bertelsmann Group has built its business on the distribution of free content.... The problem of our news portals is that the advertising effect of online advertisements is still too low.... The iPad is a chance here." Manager 23: "We have also had paid-content debates before. We usually talk about that when it gets economically critical. However, as soon as the advertising market picks up speed again and you take notice that nobody is interested in paying for news, it is no longer an issue." Manager 12: "The opportunity of the iPad is... to get access to TV-advertising revenues.... You have the possibility to integrate motion pictures in the newspaper environment."

(continued)

TABLE 4.1 Overview of the News Publishers' Lock-in Situations (continued)

Description of news publishers' lock-in	Representative supporting data
Time period 5: Since 2012, resource-based lock-in • Because overall revenue had remained very low for many years, news publishers had to reduce their editorial offices by up to 70 percent, resulting in a drastic decay of content quality. • When discussing opportunities for digital paid content now, the branch highlights that the content they produce is not of a sufficient quality to differentiate the digital newspaper product from free deals, such as free news pages or apps of TV stations and magazines with similar content. *Publishing paid digital news is not a strategic alternative for the core market, as long as the digital news product fails to have a sufficient quality to differentiate it from competing free news offerings in other sub branches. Major investments in content production would be required to consider an attempt to overcome the free content culture of news consumption.*	Manager 21: "I'm fundamentally skeptical whether there could be willingness to pay for content that is accessible for free in so many portals anyhow." Manager 9: "There are no more key media in a sense that you commit yourself to a specific news brand." Manager 21: "We publish news you get at 20 different places at the same time. . . . Making this news subject to charge just doesn't match my idea of demand and supply."

developing digital media products and using digital technology for the development and commercialization of different products. While the internet did not provide the same complementarity as paper-based news products—that is, cost advantages of combining ads and journalism in one product—it allowed news publishers to reduce overall production costs, as there would have been no need for paper, print, and expensive physical distribution—together approximately 53.5% of the costs of a printed newspaper—in a purely digital newspaper product (BDZV, 2002). With such cost reductions, revenue from regular readers who paid for the printed product in 2001—approximately 30% of overall revenue—would have been sufficient to refinance the entire content production for a purely digital product. Thus, whereas a printed journalistic product was highly dependent on advertising revenue, the digitization of newspaper content allowed news publishers to produce a daily high-quality news product and to sell it profitably without ad-based subsidies to the readership for the first time.

By the end of the 1990s, the online advertising market simultaneously began to grow, although on a very low level (BDZV, 2000). This development provided another technological opportunity: the missing need to subsidize journalistic content with advertising revenue would have allowed news publishers to develop online advertising businesses in non-content environments, such as classified ad portals for jobs and real estate.

> These [classified ads] were traditional newspaper businesses, and if publishers had been a bit more creative and faithful, they could also have founded such businesses in the digital sphere. However, the publishers remained passive. (Manager 2)

Thus, this initial opportunity of separating content and advertising businesses in the digital sphere elapsed because of a cognitive lock-in that limited the news publishers' perceived scope of action: Because classical advertising markets were still very profitable in the 1990s, the news publishers did not recognize the internet either as a potentially attractive advertising market or as an appealing readership environment, as they directly compared the online reading experience with the print experience (see Table 4.1 for a description of this cognitive lock-in).

> Haptic is very important. You don't want an electronic device during breakfast next to your coffee cup.... It is quite superior when you have a printed page that shows you everything important at a glance. (Manager 18)

Thus, news publishers perceived major shortcomings of online news products in comparison to printed newspapers from a consumer perspective: (a) the inability to give a thematic overview (like large newspaper pages), (b) the need to sit in front of a PC, and (c) the dependence on an internet connection. In the news publishers' view, quality content needed a quality medium, and during the 1990s, paper was regarded as the higher-quality medium compared with the internet.

Although news publishers still focused on selling their printed product, they felt the need to be present on the web in order to self-advertise their printed products. In the digital sphere, they decided to be free for the readers, with the aim of strengthening newspapers' overall brand coverage in order to attract readers and increase the attractiveness of the brand for advertisers. This was the initial step for an emerging free news consumption culture in digital environments.

2001–2007: The Second Window of Opportunity: Developing a Willingness to Pay and Payment Technologies

Due to the major crisis in the advertising market in 2001, which was linked to the rupture of the dot-com bubble, news publishers initially felt vindicated about their assumption that the internet would hardly be an attractive business environment for them.

> Yes, the crisis led to a very critical attitude in the publishing houses—when managers said, "I told you so: the internet will not work for news publishers." (Manager 8)

However, the online advertising market began growing again relatively soon, while revenue in the print advertising market remained low and decreased even further (BDZV, 2004; ZAW, 2004). Consequently, news publishers started to recognize that it probably was disadvantageous to reproduce their sole strategic focus on the print business and to stay passive in the digital environment. As more and more start-ups began to establish digital business models with which they successfully competed against the news publishers' print advertising businesses, publishing houses gave up their hesitant position, resulting in a massive expansion of their online news presence.

> The most important part of our restructuring was the expansion of our online news portal between 2003 and 2005. (Manager 21)

The primary goal in reaction to the major crisis in 2001 was now to regain the lost digital advertising business. Although paid content was discussed as a strategic alternative to serving the emerging free news consumption culture, it was seen as hindering the potential for developing flourishing advertising businesses in the digital sphere that would support their established business model.

> Of course, you can earn money through the reader... but I think this is not a very intelligent approach.... Perhaps, we could develop entirely different revenue streams [in the ad market]—like Google shows us. (Manager 2)

In direct consequence of this free content approach to increase revenue in the advertising market, news publishers renounced the emerging technological scope of action to create digital willingness to pay for attractive journalistic products in the readership market. Indeed, the news websites were massively expanded but largely remained free for users. Furthermore, news publishers did not make any assiduous efforts to develop digital payment solutions, such as today's famous PayPal.

> Publishers are per se no adequate innovation environment for such technologies.... Our competence is to produce content. (Manager 26)

With the decision to focus instead on regaining the digital advertising business than establishing a paid content culture, the level of path dependence changed from cognitive to normative (see Table 4.1 for a description of this normative lock-in). Due to this normative lock-in, digital technologies that build and skim off willingness to pay remained unused and undeveloped. During this time period, the free content culture of digital news consumption became widely established as a direct consequence of utilizing digital technologies solely for strengthening the advertising business.

2008–2009: The Inability to Change

When a second major advertising crisis hit the newspaper industry in 2008, news publishers intensified their paid content discussions. Although news publishers recognized that advertising revenue would not be sufficient to produce high-quality content in the medium or even long term, a shift to digital paid content products was nearly impossible at this time, as news publishers perceived themselves as exposed to a now established free content culture.

> Sadly, we have accepted over time that a general free content culture has become established on the internet, which is hardly reversible. (Manager 28, trade press interview)

As the news publishers did not dispose of resources to adequately react to implement paid content business models in this free content culture, they were locked in due to their resource base in 2008, such that their scope of action did not contain any exploitable strategic alternative. However, this resource-based lock-in (see Table 4.1 for a description) was a direct consequence of the prior focus on providing digital content for free instead of establishing a paid content culture.

2010–2011: The Third Window of Opportunity: Utilizing Mobile and Tablet Systems for Developing Differentiated Products and Establishing a Digital Paid Content Culture

When tablet PCs and smartphones arrived and were widely used for media consumption, the scope of action reopened and news publishers received a third major opportunity to move away from their traditional business model and shift their strategic focus from the advertising to the readership market: As mobile operating systems like Google Android or Apple's iOS created completely new environments for media consumption and relied on easy-to-use technologies for accessing and paying for media content, news publishers faced the strategic alternative of establishing a new paid content culture for tablet or mobile systems. With the introduction of tablet systems and smartphones, such as Apple's iPhone and iPad, many of the limiting technological restrictions that news publishers were exposed to in terms of paid content were dissolved. First, mobile operating systems like iOS or Android introduced app systems and, thus, the possibility of creating closed product environments on computing systems. Second, such tablet systems relied on technologies that eased the process of paying for content. Third, users became independent from sitting in front of a desktop computer. Instead, they were able to consume news content at different places—for instance, on a sofa, in a train, or in the park. Additionally, tablet PCs altered the media consumption process: Users no longer needed a mouse or keyboard but could browse the news with their fingers, resembling the former product concept of printed newspapers. However, although most of the observed news publishers released paid content apps, nearly all of them retained free access to their news pages, offering similar, or, in most cases even the same, content. When the advertising market started to recover from its second crisis in 2010, news publishers mainly intended to expand their brands by means of a co-existence of app and online offerings. Because users could (and can still) access the news publishers' web pages through browsers on their smartphones and tablet PCs, the use of paid news apps remained low. Ultimately, it was another normative decision (see Table 4.1 for a description of the second normative lock-in) to keep free content news pages accessible: because the advertising market recovered again, news publishers aimed to utilize these new digital technologies for stimulating the digital advertising business once more.

> Let's assume that... tablet PCs become established as highly attractive advertising medium. Then, the question arises why you should restrict the coverage of an app by subscription fees.... The question regarding the digital business model [paid or free] is still unanswered. (Manager 29, trade press interview)

Since 2012: The Re-Emergence of a Resource-Based Lock-In

The co-existence of paid but rarely purchased apps and heavily used free online news pages undermined the development of a paid content culture as a strategic alternative to the established business model. In addition, news publishers also became more and more skeptical about whether their content still had sufficient quality to justify paid content approaches. Due to the necessary savings during the decade of crisis, editorial offices had to drastically reduce their scope of service. Consequently, news publishers now find themselves confronted with numerous competitors from other industry segments, such as magazines or TV stations, which offer similar content in the digital sphere in addition to their original businesses.

> You decide to reduce the costs—it does not work; you decide to offer free content on the internet—does not work either; you invent a paid newspaper app for the iPad—could work, but then, the ARD [public television] releases a free app and you dramatically lose coverage. (Manager 28, keynote speech, media trade show)

Thus, news publishers are still bound to their ad-based business model that requires them to offer digital news for free. Overcoming this entrapment and opening up the scope of action again would now require news publishers to make major investments in their resource base—that is, in differentiating their content as qualitatively superior. Considering the low revenue generated in both the advertising and readership markets, such investments currently seem difficult to realize (see Table 4.1 for a description of the second resource-based lock-in).

DISCUSSION: LOCK-IN AS A PROCESS

Implications for Research

According to path dependence theory, firms that reproduce a strategic path do not perceive any alternative to their strategic pattern and, therefore, feel constrained in their possibilities to achieve strategic change (Sydow et al., 2009). However, this circumstance is not equivalent to not having strategic options available. Only in case of resource-based lock-ins do affected firms not have a scope of action available to deviate from strategic paths (Rothmann & Koch, 2014). By focusing our study on the lock-in stage of path dependence by extending beyond the emergence of strategic paths that much of the literature on path dependence has illuminated (Dobusch & Schüßler, 2013), our results show that path-dependent firms may be exposed to a series of windows of opportunity that remain unexploited even in times of crisis. Thus, lock-ins can continuously alternate between resource-based, cognitive, and normative forms, as new technological opportunities emerge and disappear.

Furthermore, we show how lock-ins develop over time and inhibit the exploitation of technological opportunities. Whereas firms may initially fail to recognize these strategic options in a cognitive lock-in or reject them in a normative lock-in, such lock-ins may become resource-based over time. In particular, even if news publishers had actively attempted to leave the strategic path of renouncing revenue from selling produced content online during the second major industry crisis, the challenge to succeed in this aim would have been much higher compared to the first years of product digitization in the newspaper industry, such that news publishers had no sufficient corridor of options left. Thus, our findings offer a process-based understanding of the lock-in stage (Schreyögg & Sydow, 2011; Sydow, Windeler, Müller-Seitz, & Lange 2012) that extends beyond static ascriptions of reasons for leaving technological opportunities unexploited and emphasizes the alternation of forms of lock-in over time in which the strategic potential of technological opportunities may vanish over time and windows of opportunity are eventually closed.

Implications for Policy and Practice

Our results also suggest that firms in resource-based lock-ins may not be trapped without having any alternative for an unlimited period of time. Instead, their scope of action may be broadened by the emergence of further technological opportunities. Technological advances such as smartphones and tablet PCs opened up yet another window of opportunity for news publishers. Thus, whereas the previous literature draws a rather pessimistic picture of firms' possibility of exiting the lock-in stage (e.g., Sydow et al., 2009), our process-based analysis indicates that even firms in resource-based lock-ins may not entirely be left behind without any prospects.

However, our empirical evidence also suggests that path-dependent firms are hardly able to exploit emerging technological opportunities on their own. In particular, based on our analysis, it seems that the lock-in alternates back to other forms, a normative lock-in in our case, such that path-dependent firms seem to miss emerging technological opportunities again. However, because firms can overcome both cognitive and normative forms of lock-in by unfreezing and changing the underlying cognitive and normative patterns with the help of external interventions (Sydow et al., 2009), the process-based understanding of the lock-in stage that we advance in this chapter suggests that firms have the opportunity to escape path dependence with external help if they are not trapped in a resource-based lock-in.

Limitations and Directions for Future Research

Our research results should be contemplated in light of a number of limitations. The results are drawn from a qualitative case study and are therefore by no means statistically generalizable. However, this is not the aim of qualitative research. Instead, this methodology facilitates rich empirical insights into complex social phenomena to gain novel theoretical findings through *analytical* generalization (Yin, 2014). In doing so, we unveiled those forms of lock-in that were dominant in each period. However, lock-ins hardly appear in pure forms but instead often overlap in a way that one form dominates the other (Rothmann & Koch, 2014). Taking into consideration the interplay between dominant and latent forms of lock-in could generate insightful contributions to extending our knowledge of the evolution of the scope of action in lock-ins, which we regard as a fruitful avenue for future research. Furthermore, although the joint lock-in of firms in this industry suggests an embedded single-case design, this design does not facilitate considerations of firm-idiosyncratic differences. For instance, the ownership structure of news publishers differs considerably, which may result in diverging abilities to leverage digital technology once an opportunity is recognized (Schäfferling, Wagner, & Becker, 2012). The same holds true for cross-national differences between individual firms. Nevertheless, the cognitive and normative barriers to exploiting such opportunities became manifest in an industry-level discourse (Phillips & Hardy, 2002), which may explain why all firms in this industry—even worldwide—were bound to similar restrictions (Gilbert, 2005; Koch, 2008; Oestreicher-Singer & Zalmanson, 2013). We regard this phenomenon as a fruitful direction for further research.

CONCLUSION

Our research question was the following: *How does the scope of action evolve in lock-ins?* Based on path dependence theory, we conducted a case study of the German newspaper industry and found that the range of potential strategic options continuously changes with the emergence of new technological opportunities. However, cognitive and normative restrictions inhibit the perception of new strategic options as viable alternatives until windows of opportunity close. Nevertheless, the emergence of new technological opportunities may eventually re-broaden the narrowed scope of action and provide further opportunities to escape the established strategic pattern.

Our chapter offers several theoretical and practical contributions to the literature on path dependence. First, our study complements prior research that contributed valuable insights into the emergence of paths (Dobusch & Schüßler, 2013) by focusing our research efforts on the lock-in phase of path dependence that has received less research attention so far (Rothmann & Koch, 2014). In particular, our examination of the evolution of the scope of action in lock-ins indicates that the scope of action continuously alternates between forms of lock-in that can be overcome and forms that are hardly escapable.

Second, our longitudinal analysis of a lock-in provides a process perspective on the evolution of the scope of action. In particular, our study extends beyond much of the extant literature that statically ascribes reasons for the inability of path-dependent firms to exploit emerging technological opportunities (e.g., Sydow et al., 2009) by presenting empirical evidence that these reasons alternate over time. Thus, our process perspective on the lock-in phase provides a more comprehensive answer to the question of why some firms are unable to exploit technological opportunities.

Third, in practical terms, the process perspective on the lock-in stage also points toward ways to overcome path dependence. In particular, our analysis suggests that technological changes provide various windows of opportunity over time that remain unexploited mainly for cognitive and normative reasons. Although path-dependent firms seem unable to overcome these cognitive and normative constraints on their own, they may be able to do so with the help of external interventions (Sydow et al., 2009) and, therefore, regain their ability to exploit emerging technological opportunities. Furthermore, as our findings indicate that windows of opportunity seem to close only temporarily, even firms in resource-based lock-ins may receive further opportunities to leave the path. Thus, although the news publishers have already missed a number of windows of opportunity, they may still have a promising future as new technological opportunities emerge—even in the digital sphere.

ACKNOWLEDGMENTS

The authors thank the participants of the 30th EGOS Colloquium, the 22nd European Conference on Information Systems, and the 6th International

Symposium on Process Organization Studies for insightful comments on earlier versions of the manuscript. A previous, less developed version of the paper has been included in the Proceedings of the 22nd European Conference on Information Systems.

REFERENCES

Arthur, W. B. (1989). Competing technologies, increasing returns, and lock-in by historical events. *Economic Journal, 99*, 116–131.

Arthur, W. B. (1994). *Increasing returns and path dependence in the economy*. Ann Arbor, MA: University of Michigan Press.

BDZV. (2000). *Zeitungen 2000*. Berlin, DE: Zeitungsverlag Service GmbH.

BDZV. (2002). *Zeitungen 2002*. Berlin, DE: Zeitungsverlag Service GmbH.

BDZV. (2003). *Zeitungen 2003*. Berlin, DE: Zeitungsverlag Service GmbH.

BDZV. (2004). *Zeitungen 2004*. Berlin, DE: Zeitungsverlag Service GmbH.

BDZV. (2005). *Zeitungen 2005*. Berlin, DE: Zeitungsverlag Service GmbH.

Blum, R., Bonfadelli, H., Imhof, K., & Jarren, O. (2011). *Krise der Leuchttürme öffentlicher Kommunikation: Vergangenheit und Zukunft der Qualitätsmedien*. Wiesbaden, DE: VS Verlag für Sozialwissenschaften.

Burgelman, R. A. (1991). Intraorganizational ecology of strategy-making and organizational adaptation: Theory and field research. *Organization Science, 2*(3), 239–262.

Corbin, J., & Strauss, A. (1990). Grounded theory research: Procedures, canons, and evaluative criteria. *Qualitative Sociology, 13*(1), 3–21.

Danneels, E. (2003). Tight-loose coupling with customers: The enactment of customer orientation. *Strategic Management Journal, 24*(6), 559–576.

David, P. A. (1985). Clio and the economics of QWERTY. *American Economic Review, 75*(2), 332–337.

Dobusch, L., & Kapeller, J. (2013). Breaking new paths: Theory and method in path dependence research. *Schmalenbach Business Review, 65*(3), 288–311.

Dobusch, L., & Schüßler, E. (2013). Theorizing path dependence: A review of positive feedback mechanisms in technology markets, regional clusters, and organizations. *Industrial and Corporate Change, 22*(5), 617–647.

Dovifat, E. (1967). *Zeitungslehre II*. Berlin: E. de Gruyter.

Drnevich, P. L., & Croson, D. C. (2013). Information technology and business-level strategy: Toward an integrated theoretical perspective. *MIS Quarterly, 37*(2), 483–509.

Geertz, C. (1973). Thick description: Toward an interpretative theory of culture. In C. Geertz (Ed.), *The interpretation of cultures: Selected essays* (pp. 3–30). New York, NY: Basic Books.

Gilbert, C. G. (2005). Unbundling the structure of inertia: Resource versus routine rigidity. *Academy of Management Journal, 48*(5), 741–763.

Hannan, M. T., & Freeman, J. (1977). The population ecology of organizations. *American Journal of Sociology, 82*(5), 929–964.

Hoskins, C. (2004). *Media economics*. Thousand Oaks, CA: Sage.

IVW. (2017). *Quartalsauflagen Tageszeitungen*. Retrieved from http://www.ivw.eu/aw/print/qa

Koch, J. (2008). Strategic paths and media management: A path dependency analysis of the German newspaper branch of high quality journalism. *Schmalenbach Business Review, 60*(1), 51–74.

Koch, J. (2011). Inscribed strategies: Exploring the organizational nature of strategic lock-in. *Organization Studies, 32*(3), 337–363.

Kunow, K., Gersch, M., & Koch, J. (2013). Temporary incompetence as a path-breaking strategy: Two major record companies' efforts to escape their competence lock-in. *Jahrbuch Strategisches Kompetenz Management, 6*, 9–33.

Langley, A. (1999). Strategies for theorizing from process data. *Academy of Management Review, 24*(4), 691–710.

Leonard-Barton, D. (1992). Core capabilities and core rigidities: A paradox in managing new product development. *Strategic Management Journal, 13*(S1), 111–125.

Lincoln, Y. S., & Guba, E. G. (1985). *Naturalistic inquiry*. Newbury Park, CA: Sage.

Lopes, A. B., & Galletta, D. F. (2006). Consumer perceptions and willingness to pay for intrinsically motivated online content. *Journal of Management Information Systems, 23*(2), 203–231.

Mahoney, J. (2000). Path dependence in historical sociology. *Theory and Society, 29*(4), 507–548.

March, J. G. (1991). Exploration and exploitation in organizational learning. *Organization Science, 2*(1), 71–87.

Miller, D. (1993). The architecture of simplicity. *Academy of Management Review, 18*(1), 116–138.

Oestreicher-Singer, G., & Zalmanson, L. (2013). Content or community? A digital business strategy for content providers in the social age. *MIS Quarterly, 37*(2), 591–616.

Pentland, B. T., Feldman, M. S., Becker, M. C., & Liu, P. (2012). Dynamics of organizational routines: A generative model. *Journal of Management Studies, 49*(8), 1484–1508.

Phillips, N., & Hardy, C. (2002). *Discourse Analysis: Investigating the Processes of Social Construction*. Thousand Oaks, CA: Sage.

Pierson, P. (2000). Increasing returns, path dependence, and the study of politics. *American Political Science Review, 94*(2), 251–267.

Rothmann, W., & Koch, J. (2014). Creativity in strategic lock-ins: The newspaper industry and the digital revolution. *Technological Forecasting and Social Change, 83*(3), 66–83.

Schäfferling, A., Wagner, H.-T., & Becker, J. (2012, June). Exploring the relation between firm ownership and IT capability. *Paper presented at the 20th European Conference on Information Systems*, Barcelona, Spain.

Schmidt, T., & Braun, T. (2015). When cospecialization leads to rigidity: Path dependence in successful strategic networks. *Schmalenbach Business Review, 67*, 489–515.

Schreyögg, G., & Sydow, J. (2011). Organizational path dependence: A process view. *Organization Studies, 32*(3), 321–335.

Schreyögg, G., Sydow, J., & Holtmann, P. (2011). How history matters in organisations: The case of path dependence. *Management and Organizational History, 6*(1), 81–100.

Shapiro, C., & Varian, H. R. (1999). *Information rules: A strategic guide to the network economy.* Boston, MA: Harvard Business School Press.

Staw, B. M. (1981). The escalation of commitment to a course of action. *Academy of Management Review, 6*(4), 577–587.

Stinchcombe, A. (1965). Social structures and organizations. In J. G. March (Ed.), *Handbook of Organizations* (pp. 142–193). Chicago, IL: Rand McNally.

Sydow, J., Schreyögg, G., & Koch, J. (2009). Organizational path dependence: Opening the black box. *Academy of Management Review, 34*(4), 689–709.

Sydow, J., Windeler, A., Müller-Seitz, G., & Lange, K. (2012b). Path constitution analysis: A methodology for understanding path dependence and path creation. *Business Research, 5*(2), 155–176.

Sydow, J., Windeler, A., Schubert, C., & Möllering, G. (2012a). Organizing R&D consortia for path creation and extension: The case of semiconductor manufacturing technologies. *Organization Studies, 33*(7), 907–936.

Tripsas, M. & Gavetti, G. (2000). Capabilities, cognition, and inertia: Evidence from digital imaging. *Strategic Management Journal, 21*(10/11), 1147–1161.

Tyre, M. J., & Orlikowski, W. J. (1994). Windows of opportunity: Temporal patterns of technological adaptation in organizations. *Organization Science, 5*(1), 98–118.

Van de Ven, A. H. (1992). Suggestions for studying strategy process: A research note. *Strategic Management Journal, 13*(S1), 169–188.

Vergne, J.-P., & Durand, R. (2011). The path of most persistence: An evolutionary perspective on path dependence and dynamic capabilities. *Organization Studies, 32*(3), 365–382.

Wenzel, M. (2015). Path dependence and the stabilization of strategic premises: How the funeral industry buries itself. *Business Research, 8*(1), 265–299.

Wenzel, M., Senf, N. N., & Koch, J. (2016). Examining complex phenomena with qualitative research methods: An examination of strategic innovation trajectories in haute cuisine. In: E. S. C. Berger & A. Kuckertz (Eds.), *Complexity in entrepreneurship, innovation and technology research: Applications of emergent and neglected methods* (pp. 163–182). Cham, Switzerland: Springer.

Wohlgemuth, V., & Wenzel, M. (2016). Dynamic capabilities and routinization. *Journal of Business Research, 69*(5), 1944–1948.

Yin, R. K. (2014). *Case study research: Design and methods* (5th ed.). Thousand Oaks, CA: Sage.

Zajac, E. J., Kraatz, M. S., & Bresser, R. K. F. (2000). Modeling the dynamics of strategic fit: A normative approach to strategic change. *Strategic Management Journal, 21*(4), 429–453.

ZAW. (2004). *Werbung in Deutschland.* Berlin, Germany: Author.

CHAPTER 5

RESOLVING THE PARADOX OF TECHNOLOGICAL INNOVATION

The Role of Policy and Practice in Protecting Employee Well-Being

Stacie Furst-Holloway and Daniele A. Bologna
University of Cincinnati

ABSTRACT

Technological innovation has dramatically reshaped the modern workplace. When, how, and where employees work is no longer a reliable certainty. Indeed, the use of mobile communications technologies (MCTs) has blurred the physical, temporal, and psychological boundaries between the work and nonwork domains for many employees. However, evidence regarding the impact of MCT use on employee and organizational outcomes is nebulous at best, with some studies linking MCT use to positive outcomes (e.g., work–life satisfaction, engagement) and other studies linking MCT use to less desirable outcomes (e.g., burnout, stress, distractions). Using data from our own research as well as the broader academic and practitioner literatures and

Technological Innovation Networks, pages 99–121

popular press, in this chapter we address this discrepancy in the literature by (a) illustrating the specific contradictions—or paradoxes—associated with MCTs and their impact on both work outcomes and employee well-being; (b) describing how organizational norms, policies, and expectations regarding MCT use can contribute to those conflicting outcomes; and (c) discussing the implications for organizational as well as national policy and practice that can promote more consistently positive outcomes from using MCTs to manage work and life demands.

Technological innovations, such as the development of laptops, tablets, smartphones, and smart watches, have dramatically reshaped how, when, and where work is accomplished in organizations across the globe (Leung, 2011). In 2008, 87% of U.S. workers reported that they engage in work-related activities while at home for at least some part of the work week. Further, 17.2 million employees self-identified as telecommuters, up from 12.4 million in 2006 (WorldatWork, 2009). Indeed, the use of mobile communications technologies (MCTs) now blurs the physical, temporal, and psychological boundaries between the work and nonwork domains for many employees (Haeger & Lingham, 2014).

Research suggests that the integration afforded by MCTs benefits employees by providing greater convenience and flexibility, boosting productivity, and making work more interesting (Jarvenpaa, Lang, & Tuunainen, 2005). The demand for greater flexibility and new ways of working derives, in part, from the changing nature of the workforce, not just in the U.S. but throughout the developing world. As an example, in 2012 women spent more time on paid work and significantly less time on housework work than they did in 1965. For men, these patterns are reversed. In fact, among U.S. families with a working mother today, one in five households with small children has a stay-at-home dad (Rouse, 2014). Issues of work–life integration now expand beyond parent–child and dual career relationships to include issues of eldercare as well. As the baby boom generation ages, more employees experience "sandwich effects" of having to manage childcare responsibilities as well as caring for elderly parents. As the labor force becomes more diverse and gender roles continue to evolve, organizations must adapt and find new ways to support employees who must balance myriad work and nonwork demands. To this end, calls have come from government offices, including the White House, that encourage organizations to consider adopting flexible practices that can "potentially boost productivity, improve morale, and benefit the economy" (Rouse, 2014, ¶7).

While the benefits of flexibility are clear, we caution that there is simultaneously a "dark side" (Ran, 2013) to MCT use that arises from having around-the-clock (24/7) access to work and the inability to detach, both of which can lead to greater stress, work–home conflict, and burnout (Sarker, Xiao, Sarker, & Ahuja, 2012). A clear tension therefore exists with technology-enabled work: The same capabilities that boost employee flexibility and productivity can also threaten their well-being and performance (Jarvenpaa & Lang, 2005). Despite burgeoning evidence regarding these paradoxical outcomes, the trend toward greater technology access and flexibility continues somewhat unabated. Far less attention has been paid to the organizational and societal practices and policies that can preserve productivity and performance while also protecting employees from the psychological and physiological effects of anywhere, anytime work.

Drawing upon data from our own research as well as the broader academic and practitioner literatures and popular press, the purpose of this chapter is to address this issue by (a) illustrating the specific contradictions—or paradoxes—associated with MCTs and their impact on both work outcomes and employee well-being; (b) describing how organizational norms and policies regarding MCT use can contribute to those conflicting outcomes; and (c) discussing the implications for organizational and national policy and practice that can promote more consistently positive outcomes from using MCTs to manage work and life demands.

THE PARADOX OF MOBILE TECHNOLOGY

A broad literature identifies the benefits and drawbacks associated with using MCTs to accomplish work in the virtual, digital age (Jarvenpaa et al., 2005). Our own research, supported by the Society for Human Resource Management (SHRM) Foundation, affirms many of the same—as well as a few new—conflicting outcomes, which we label "paradoxes of technology." As shown in Table 5.1, we specifically identified a set of seven paradoxes that underlie the utilization of MCTs. We describe each of these paradoxes more fully in the section that follows and then discuss the implications of these paradoxes for practice and policy.

Paradox #1: Accessibility Versus Being Accessible

Interview and survey data from hundreds of employees and HR professionals representing a range of industries, job functions, and positions of authority suggest that *accessibility to work* and *the ability to stay connected to others* are the biggest benefits afforded by MCTs. Consider the following quotes from employees our research team has interviewed:

> If there is some kind of emergency and I need to do something, I do have technology with me and I can use it to get work done if I had to someplace else.

TABLE 5.1 Paradoxes of Technology

Paradox	Benefit	Drawback
1. Accessibility versus Being Accessible	Accessing work from anywhere at any time allows you to respond and work while not in the office so that you can attend to other things or simply work when and where is most comfortable for you.	Ability to be contacted at all hours and having to engage in work when you are in another domain . . . so you never can fully detach from work and it can interfere with other roles.
2. Controlling versus Losing Control of Boundaries	The flexibility afforded by MCTs provides employees with a greater sense of control over their work, which increases engagement, commitment, and satisfaction.	When accessible to others, employees may feel pressure to respond to emails, texts, or messages. This can diminish perceptions of control, motivation, and well-being.
3. Managing Workload versus Creating More Work	Checking email before coming in to gauge what priorities might be and to stay on top of things when out of the office so that when you return your work is more manageable	As a result of checking in or simply because technology lets you, the volume of work goes up—threatening the manageability of your workload.
4. Costs and Benefits of Technology	Technology allows you to work with greater speed and to save costs in terms of commute, paper, and so on.	Costs more to keep technology up to date; have to learn new software and deal with glitches that often come up.
5. Using Technology to Manage Demands and Recharge versus Distract	Using MCTs to manage non-work demands when needed, allowing employees to refocus on work; technology as a quick, refreshing escape to recharge during work day	Propensity for employees to not self-regulate and become distracted at work by access to information, news, and others via using MCTs.
6. Personal Control versus Coordination with Others	Flexibility affords individuals a lot of choice in terms of how and when they can structure their work to maximize productivity.	As individuals schedule their own work time, it increases coordination challenges for interdependent others.
7. Breadth versus Depth of Relationships	Ability to build expansive networks via a multitude of communication channels.	Networks are expansive, but ties within those networks can be quite weak.

> (Technology) makes things quicker and easier. . . . Having access at home. I feel more comfortable knowing that I can check in on staff.

Being able to access work while off hours or offsite provides employees with greater flexibility in balancing work and nonwork demands. Employees can be available to respond to emails, take calls, instant message a colleague, or literally "peek" into an office or manufacturing system to keep working without constraints on time or location. Yet, for many employees, these benefits are offset by a corresponding set of challenges associated with being accessible to others anywhere, anytime. Consider the following quotes:

> Well, in my position probably the biggest drawback [of technology] is people will text me on Friday night just to complain about something. You know, it's not necessarily a work-related issue. So because they do have access to me 24/7 sometimes that can be a drawback for me.

> It's just a quick and reliable way to keep up-to-date on everything. I think the flip side of that is that it can be abused internally where people assume you are always connected and always available.

As these quotes illustrate, the downside of connectivity is the inability for employees to fully distance themselves from work when MCTs are so pervasive. In fact, several participants said they avoided purchasing a mobile device because they wanted to protect the boundary between work and home. Others who owned mobile devices set limits on when they would be willing to access a mobile device to prevent unwanted intrusions. These tactics included turning off mobile work devices while at home or restricting access during meals or family events.

Paradox #2: Controlling Versus Losing Control of Boundaries

The tension employees experience between having access and being accessible raises a slightly different but related paradox: control. Specifically, our respondents noted that an implicit benefit of using MCTs to enjoy greater flexibility is having greater perceived job control—that is, being able to control where, when, and how work is accomplished. Consistent with the tenets of self-determination theory (Deci & Ryan, 2012), most individuals desire some degree of autonomy in their lives, including (for many) how and when they work. When that condition is met, employees are more likely to be intrinsically motivated to engage in work and to exhibit enhanced performance, persistence, and creativity. It is therefore understandable that MCT users report higher levels of job satisfaction and engagement when technology provides them with more control over their work, including how they manage the work and nonwork boundaries.

However, the same technologies that afford employees this sense of control can also undermine control by setting up expectations to be available and responsive to others at work regardless of time and location. The following quote illustrates this emerging phenomenon:

> I always know what's going on even if I'm not at work, especially because I have emails linked up to my personal stuff. You know, it could be 10:00 at night and I haven't been at work for six hours, but I get email and I'm tempted to read it and then I've got to—half the time I end up working because it's something I feel like I need to do.

The behaviors being described in this quote, as well as the sense of obligation to respond, are increasingly common. A survey of 1,000 workers by Good Technology found that 68% of respondents checked work e-mail before 8:00 a.m., 50% checked it while in bed, and 38% regularly did so during dinner (Good, 2012). Similarly, a study by the American Psychological Association (2013) revealed that 44% of working adults surveyed admitted checking work email daily while on vacation, with about 10% checking in on an hourly basis. The feelings of obligation to be responsive to others, even while on vacation, can threaten perceptions of control and undermine both intrinsic motivation and employee wellness. Yet this has become such a prevalent experience that scholars and practitioners have developed new terms, like technology-related pressure (Harris, Marett, & Harris, 2011), telepressure (Barber & Santuzzi, 2015), and addiction to technology (Turel, Serenko, & Bontis, 2011) to describe its occurrence. Further, research suggests that the pressure to respond (distinct from actually responding) increases stress, work–family conflict, and burnout, and it decreases employees' focus on their work (Barber & Santuzzi, 2015; Harris et al., 2011).

Paradox #3: Managing Workload Versus Creating More Work

In addition to issues of accessibility and control, another paradox we identified relates to the amount of work that results from reliance on MCTs. Several individuals, for instance, discussed using a smartphone or laptop to access voicemails and emails off hours or while away from work not necessarily to "do" work but to either stay on top of things or to prepare for work that awaits when they get to the office. In this sense, technology helps them to prepare mentally and physically for the work that needs to be done. Consider the following quote from a marketing director who described checking work emails that had been downloaded to his smartphone between the time he left work at the end of the day and returned the next morning:

> I know in the morning what I'm walking in to the office instead of "I hope I don't have 800 unanswered e-mails because there was some catastrophe that

> I wasn't alerted to." I already know exactly what's going on in both realms of my life at all times.

Research and the popular press are replete with examples of MCT users who similarly use technology to "check in" while on vacation in order to respond to information requests or have quick conversations with colleagues so that when they return to the office after an absence, work will not be backlogged. Technology enables these employees to manage their workload more efficiently. In contrast, other employees remarked that by accessing this information and knowing what work is there and waiting, they seem to create more work for themselves. A few quotes from our study participants illustrate this effect:

> Well, obviously you can take it anywhere and get it done so I mean it creates more work, but it's very convenient to be able to be sitting on the soccer field with a laptop.
>
> Well, it allows you to do more work because everything's at your fingertips. You can get things done more efficiently, but people expect more out of you, so they give you more work to do.

As these quotes suggest, when colleagues and customers can reach you around the clock, employees may increase their workload in the form of answering questions, being pulled into group email conversations, and being asked to comment on other happenings in the office. Fenner and Renn (2009) refer to this as *technology-assisted supplemental work* (TASW).

Paradox #4: Costs Versus Benefits

A fourth paradox relates to the perceived costs and benefits associated with the adoption of MCTs to engage in work. Specifically, our findings suggest that many employees associate mobile technology with increased efficiencies at work as a result of (a) shifting from paper to digital formats, (b) using technology to communicate with multiple people at once, and (c) reducing time and travel costs. The following quotes illustrate some of the positive perceptions relating to technological advancements:

> I really depend on [technology] a lot. I get on a computer and see what all the equipment is doing. I don't have to actually go through the facility and look at it and look at every detail... I'm on the computer.
>
> There are sometimes one-hour meetings that will require me to drive, spend the gas to drive just for that one meeting. And I will use technology, as an

> excuse "Hey, can I just call in? Can I just Skype this?" And I use it as a reason to not have to go.

Clearly, technology allows employees to work more efficiently and to save time and costs by switching from manual processes and in-person work to mobile, virtual, and electronic spaces. Yet technology also has a cost. Equipment must be purchased and maintained and time must be taken to set up, learn, and navigate new systems. Moreover, technology is not always reliable. Glitches or viruses may pop up that can slow down or completely shut down work. Internet access and cell phone service, particularly when traveling, may be spotty. In each of these cases, an overreliance on technology to accomplish work means a heightened sense of dependence on the technology to be available and functional. This is not always the case, as these quotes illustrate:

> We have had network issues before. So then we can't have... actually there was one time I think the whole shared drive that everyone has access to was down for at least a day. If there are any technological issues then we can't get a lot of work done.

> [Technology] creates a lot of efficiencies but it creates different kinds of work. Like now instead of writing a 20-page paper to convey everything that I have discovered on a project I can take two hours and generate a 25-slide PowerPoint.

Glitches and downtime associated with other technology-related problems can create a fair amount of stress for employees who rely on these tools to stay connected and productive. Failing to back up data or to identify contingency plans when technology fails can exacerbate these effects and drive up some of the hidden costs of technological dependence.

Paradox #5: Recharging Versus Distraction

A fifth paradox we identified may be the most challenging for organizations to navigate: the utilization of MCTs to provide a momentary break from the rigors of work versus the potential for MCTs to distract employees from work. Much has been written on the concept of cyberloafing, when employees access the company internet during work hours to engage in nonwork tasks (Henle & Blanchard, 2008; Lim, 2002). Some reports suggest that employees spend up to 2 hours per 8-hour work day on technology-enabled nonwork activities, such as responding to emails on a personal email account, checking friends' Facebook updates, reading sports or news, shopping, and paying bills online (Henle, Kohut, & Booth, 2009; Vitak, Crouse, & LaRose, 2011). As a result of or in anticipation of these behaviors, many companies establish policies that prohibit the use of the Internet for anything other than work during normal operating hours. Several comments from employees we studied underscore why such policies may be needed, as even those with the best of intentions to simply "log in" for a few moments to take a mental break can easily become distracted. For instance:

> I go to Yahoo a lot and I'll read one article and then I'll go back to the main page, and instead of going back to what I was working on, I'll see another article and I'm like "Oh crud, I got to read that one." And then sometimes it will be a little bit of a snowball effect where I'll read a sports article and then, "Oh, I haven't checked ESPN today, maybe I should go do that next before I get back to my activity that I'm working on."

Without strict guidelines or some level of self-regulation on the employee's part, access to the Internet or other resources available via MCTs can impede productivity. Yet again, the picture is far more complex. We heard from many employees who report clear benefits from being able to access technology or to go online. For them, the ability to manage important non-work demands (e.g., accessing the internet to pay a bill, register for a class, or check in with a family member) was important. Being able to check some things off their to-do list while at work alleviated some pressure and allowed them to return their energies and focus to work.

Beyond helping to manage these other demands, technology-related breaks can also provide a mental break that some people feel they need (Kim, Bologna, Furst-Holloway, Hollensbe, Masterson, & Sprinkle, 2014). The following quote describes this strategy.

> Every now and then you have to let your mind have a break from what you're working on. I think it can be counterproductive if you kind of abuse (technology) and take too much time, and you're looking at getting too involved in things online, so it's kind of getting that separation of moderation I guess.

Finally, when it comes to technology-related policies in the workplace, it is worth noting that employs pick up on intended or unintended signals from their employer when strict rules regarding internet access are or are not in play. Respondents we interviewed who had experienced both types of environments suggested that restrictions on use can signal that the company mistrusts employees, leaving them to feel less supported by and trusting of their employer. The following comment illustrates these perceptions:

> I would like to see HR... say "We recognize you all have cell phones in your pockets. Let's use them respectfully. We are not going to chastise you if you use social media. Please do it professionally. Here are some guidelines." Instead of just saying, "[Technology use] is forbidden."

In these instances, companies inadvertently may set up situations where employees look for other means to recharge (e.g., longer lunch breaks or coffee break) or they simply skirt the rules by accessing the internet via a personal device.

Paradox #6: Independence Versus Collaboration and Interdependence

Although not as salient in our own work, the broader literature identifies two other paradoxes that we believe are important to note in any discussion of mobile technology. The first describes the tension between employee flexibility and independence and the realities of interdependent work. In particular, most employees have preferred work flows—those hours of the day and even physical location when and where they work best. Being able to schedule work around those peak times, coupled with the ability to manage nonwork demands (e.g., taking care of children or an elderly parent) as needed, allows employees to experience independence and feel empowered to maximize the use of their own time. However, many employees do not work independently from others. When work is interdependent, coordination challenges arise when, at the individual level, employees have the freedom to choose when and where they work (Van Dyne, Kossek, & Lobel, 2007). Frustrations may occur if an employee working on a team-based project chooses to do his work remotely and off hours due to other obligations and is thus unresponsive to teammates during the workday. As MCTs increase opportunities for employees to carve out their own personal work schedules and workflows, the risk of such coordination challenges also increases (Kossek, Thompson, & Lautsch, 2015).

Paradox #7: Relationship Quantity Versus Quality

A final paradox concerns the depth and breadth of connections and relationships formed using mobile technologies. Certainly, technology allows individuals to expand their networks and contacts in new ways. It is quick and easy to include multiple people on emails and invitations to meetings and events, as well as to access and connect to others through various listservs, databases, and social media portals, such as LinkedIn. But, as a report from the Pew Center highlights, technology may also interfere with the quality of those communications and the depth of the relationships between formed (Hampton, Sessions, Her, & Rainie, 2009). Although this study focused on nonwork relationships, including ties within a neighborhood and from voluntary associations, the findings are consistent with the broader literature. As McPherson, Smith-Lovin, and Brashears (2006) note, the type of social ties supported by MCTs are relatively weak and geographically dispersed, not the strong, often locally based ties that tend to be a part of peoples' core discussion network. In the context of work, this suggests that employees may spend more time connecting with and communicating with remote others in a superficial way while engaging in less rich communications with colleagues in their immediate work environment. The following observations from our study participants echo these concerns:

> It's all about face-to-face with the customer but a lot of times people just email us now, which is helpful because you can get back to them quickly, but then you kind of lose the face-to-face interaction. So it's not as personal.
>
> If you're able to have a conversation with someone face to face I feel like it's the absolute most efficient way to transfer information and insure 100% understanding because so much gets lost when you go degrees of separation away from face to face.

Indeed, previous research has argued that rich and supportive environments are difficult to create in a virtual setting (Nafukho, Graham, & Muyia, 2009), reflecting another clear paradox of mobile technology use.

Summary of the Paradoxes of Technology

Taken together, the identification of these seven paradoxes, as well as others perhaps not yet recognized, suggests that the proliferation of MCTs does not come without significant costs to employee well-being and performance. Yet evidence seems to suggest that the trend toward greater reliance on technology to meet work and home demands will continue. While acknowledging the benefits afforded by MCTs, we caution that not enough attention has been paid to the physiological and psychological effects of 24/7 connectivity, the increased pace of work, and the stress associated with being responsive, available, and engaged.

Our concerns, specifically, reflect several disconcerting health trends seen throughout the developed world. A study commissioned by the Association for HR Management in International Organizations found that 58% of individuals report losing sleep over work issues (Shortland, 2005). Research from the Families and Work Institute also reveals that one in three employees in the U.S. reports being chronically overworked and that 36% of those considered highly overworked experience extreme levels of stress (Galinsky, Bond, & Hill, 2004). While nearly half of these employees feel that they are in poor health, a startling 21% have symptoms of clinical depression. Finally, the International Labor Office (2004) predicts that—around the globe—the occupational diseases of the 21st century will be heart attacks, strokes, and suicide. As we discuss in the sections that follow, organizational and social policy and practice need to consider their role in ameliorating these effects.

DISCUSSION

Impact on Organizational Policy and Practice

Given that the potential gains in productivity and flexibility associated with MCT use may put employees at risk, it is not surprising that researchers, practitioners, and policymakers have begun to investigate the state of policy and practice in governing healthy workplace behaviors (e.g., Furst-Holloway & Jones, 2011). These efforts reveal that, in general, *practice* and *policy* at the organizational level have been slow to respond. This suggests either a lack of knowledge regarding the ill-effects of pervasive technology use or perhaps a preference for short-term organizational gains in productivity over the long-term health and well-being of their employees. More realistically, we expect that most employers simply do not know how to help employees leverage MCTs in a way that provides the flexibility they desire while also maintaining opportunities to detach, refresh, and simply recharge from time spent on work. What evidence we do have suggests that formal and informal policies and practices have a tremendous amount of influence on the extent to which employees feel pressure to stay connected. Briefly, we describe some of these forces, including flexibility policies, role expectations, and organizational norms regarding technology use.

Workplace Flexibility Policies

Organizations vary considerably in terms of the policies they adopt to manage employees' use of technology, oftentimes reflecting the nature of the industry or business in which they operate. For instance, multinational companies must support the ability for globally dispersed colleagues to communicate in real time across time zones. Similarly, organizations with 24/7 operations (e.g., healthcare facilities and call centers) must have key personnel available at any time to respond to critical issues impacting a patient, customer, or operations. However, even within a given industry, variability may exist that reflects the culture and business strategy of the organization.

At one end of the spectrum are those organizations that fully embrace the flexibility afforded by mobile technology and anytime, anywhere polices. Most notable are those companies shifting to a results only work environment (ROWE). ROWE programs emphasize results over presence, allowing employees to work wherever and whenever they choose as long as they meet performance targets. Empirical evidence suggests that ROWE policies may have positive effects on the work–family interface for *some* employees but not all. For instance, in a longitudinal study of more than 608 employees participating in a ROWE program, Kelly, Moen, and Tranby (2011) found that work–family conflict diminished and work–family fit and time adequacy improved for many participants, primarily by increasing employees' schedule control. However, for those employees with high work demands, increasing schedule control had no effect on the same work–family outcomes. These findings underscore the difficulty that some employees may have in setting boundaries if demands *and* flexibility regarding when and where work is performed are both high (Blair-Loy, 2009).

In contrast to ROWE policies, other organizations develop formal flexibility policies stating which, if any, employees have the ability to work flexibly and what the conditions of those arrangements should look like. Most often these are referred to as flexible work arrangements (FWAs) and include options such as telework, job sharing, half time, or compressed work weeks. Although these arrangements are associated with higher levels of engagement and job satisfaction, they focus almost exclusively on activities that occur during normal business hours, not "off hours" work that carries over into the nonwork domain as a result of MCT-enabled access.

It is quite possible, however, that by expanding the parameters of FWAs to explicitly include consideration of off-hours work, some of the strain associated with technology-assisted supplemental work (TASW) may be eroded. For instance, although late-night use of MCTs has been shown to predict morning depletion and reduce next-day work engagement, these negative effects can be alleviated by increasing employees' sense of job control (Lanaj, Johnson, & Barnes, 2014). Thus, in the absence of FWAs, managers may wish to monitor employees' off-hours activity and be willing to adjust subsequent work hours (e.g., allowing the employee to come in late the next day) when they know an employee was connected and working late. Otherwise, they may wish to discourage off-hours work if it appears to diminish employees' focus and engagement during work hours.

Informal Norms and Expectations

While a vast literature explores the impact of formal FWAs, less attention has been paid to the informal, unspoken norms regarding technology use in organizations. Specifically, these technology-related norms represent employees' perceptions of whether individuals in the organization whose opinions they value, such as supervisors or organizational leaders, use MCTs to connect to work off hours (Richardson & Benbunan-Fich, 2011). Our own research suggests that these norms strongly influence whether employees feel pressure to stay connected. That is, when employees see their organization's leaders and perhaps their own supervisors using MCTs to engage in work off hours, they are more likely to report feeling pressure ("telepressure") to be responsive to emails, texts, or other communications regardless of time and place.

This, of course, presents a dilemma to those in supervisory positions, as the necessity for staying connected typically increases with span of control and authority. Yet, caution must be given to ensure that expectations for availability and responsiveness at all hours should not be the norm for *all* employees. To reduce these pressures, supervisors may wish to develop explicit norms regarding the division of work and family time (e.g., Lautsch, Kossek, & Eaton, 2009). Alternatively, they may adopt the practice of including in their off-hours communications language that explicitly permits employees to not respond until the next workday unless the situation is urgent. At a minimum, supervisors should be cognizant of how their own MCT use may influence employee perceptions regarding the expectations of an "ideal" employee.

In lieu of a broad company policy, organizations may instead wish to develop more specific policies at the unit or workgroup level. For instance, in a seminal study of flexible work-based interventions, Harvard's Leslie Perlow (1999) studied a group of employees from the Boston Consulting Group (BCG), who adopted a "predictable time off" policy whereby consultants identified evening and weekend periods where team members would be unreachable. This intentional time off allowed them to completely disconnect both psychologically and physically from work. Consultants using the policy exhibited lower stress, felt rejuvenated, and were more likely to report being satisfied with their jobs than those who did not use the policy. More importantly, although their weekly work hours dropped 11%, productivity levels did not diminish (Perlow & Porter, 2009). While the BCG policy holds promise, it is worth noting that the ability to disconnect may be contingent upon coworkers' and supervisors' willingness to honor that time off.

Building on Perlow's work, Van Dyne and his colleagues (2007) recommend the practice of synchronized interaction to guide behaviors that provide better balance between interdependence and individual choice. Synchronized interaction is a work practice where group members specify when interactions should occur (e.g., team meetings, one-on-one conversations) and when individual contributor work should occur. This approach gives employees uninterrupted time for individual cognitive activities that they can schedule in advance at preferred times, while assuring adequate time and availability for group collaborative activities.

Finally, organizations may also wish to consider policies on *response etiquette,* including when and how quickly employees are expected to respond to emails, texts, or instant messages. For instance, in a study of 184 participants across numerous organizations in Europe and North America, Shortland (2005) found that only 15% of respondents' organizations had guidelines regarding the speed at which employees should respond to email communications. Of those that did, guidelines ranged from same day to five days, depending upon whether the response required a simple acknowledgement or a substantive response. A more recent study by Brewer (2015) suggests that most individuals prefer response times within the 12–48 hour time range, depending again on the context. Both sets of findings suggest that most professionals do not expect immediate responses, thus reinforcing the idea that pressures to respond the same day, night, or morning may be internalized. To this end, explicit guidelines that permit delaying responses until the workday may alleviate some of this pressure.

For those unable to turn off technology, organizations may wish to invest in technology that actually prevents emails from being sent during off hours. For example, software products such as Enforced Vacation, which stops employees from emailing outside of specified hours, have increased in popularity. This software essentially holds all emails being sent to employees during specified times, such as during non-core business hours or during any nonwork time. When the workday begins, those emails are then delivered. Special features allow employees to customize their settings to allow specific communications (as indicated by key words) to filter through the screen. However, all other emails are held until the block is released.

Conclusion

The preceding list of organizational policies and practices is not exhaustive. However, it may be useful for organizational leaders and policymakers to consider when discussing the adoption of policies and practices to govern employees' use of MCTs. A singular focus on providing flexibility via MCT utilization, without consideration of the psychological and physiological effects of anywhere, anytime work, may put employees at risk for greater stress and burnout. While some leaders may be concerned that restricting off-hours access may adversely impact performance, much of the evidence points to the reverse effect: Restrictions and guidelines allow employees greater focus during work time so that productivity does not suffer. Despite these promising results, organizations may still be hesitant to build in these safeguards at the risk of cutting into profitability and competitiveness. This may explain a growing movement toward national policies directed at employee health and well-being in the midst of the technology age. We describe these efforts next.

Impact on National Policy and Practice

At the national level, concerns regarding 24/7 work shift from the balance between organizational performance and employee well-being to the balance between broader economic gains and population (physical and financial) health and well-being. Not surprisingly, a close inspection of these issues again suggests an imbalance favoring the economic side of the equation. For instance, data from the Economic Policy Institute reveal that during the great internet-age boom in productivity, in which productivity rates jumped 23% since 2000, the inflation-adjusted wages and benefits for college graduates climbed just 4%. This suggests that we are working more for less: less money, less time for leisure and family, and less time to detach from the strains of work. Accordingly, in the absence of more comprehensive employer-driven efforts to manage employee time and workload issues, broader national policies may be needed. Indeed, efforts to shape policy are beginning to emerge focusing on employee work hours as well as compensation relating to off-hours or overtime work.

Representing perhaps the most progressive move toward limiting work hours, it appears that Sweden may be moving toward a six-hour work day (Science Alert, 2015). Although not a matter of national policy at the present time, companies across the country have been adopting this practice for many years. Other organizations are beginning to follow suit in the face of evidence demonstrating better employee outcomes with no loss in productivity. Proponents of the six-hour work day plan explain that with the reduction in work hours, employees are asked to stay off social media and other distractions while at work. Meetings are also kept to a minimum. The net effect is that employees are able to stay more fully present and focused on their work as part of a reduced workday because they are free from technology-enabled distractions *during* their time on the job and because they have additional off-hours time to engage in essential or valued nonwork (personal) activities.

While Sweden may be an exception to the rule, it is also true that several other countries in Europe historically have attempted to establish work limitations to minimize the negative impact of overwork on personal well-being. In the United Kingdom (UK), for example, the 1998 Working Time Regulations limited employees to working no more than an average of 48 hours per week over a 17-week period. This effort was consistent with the European Union's Working Time Directive, which is designed to protect workers' health and safety by establishing a minimum set of standards around working hours. These standards include, but are not limited to:

- Limiting weekly working hours to no more than 48 hours on average (including overtime)
- A minimum daily rest period of 11 consecutive hours in every 24 hour period
- A minimum weekly rest period of 24 uninterrupted hours for each 7-day period, in addition to the 11 hours' daily rest
- Average working hours not exceeding 8 hours per 24-hour period.

These policies were sufficient before the proliferation of mobile technologies. But in the modern workplace, many MCT-enabled off-hours work behaviors, such as checking email at home or answer a text or IM during a child's soccer game, may go undocumented and therefore escape consideration in such work time limits. Given evidence suggesting that employees who engage in MCT-enabled work are more likely to work *above* their contracted hours (Empirica, 2000), it is therefore not surprising that media, labor rights groups, and employees have voiced concerns over employers requiring work outside of normal hours and the need to crack down on unpaid or underpaid "after-hours" TASW behaviors (Stangler, 2015).

It appears that the United States could be next to join this movement. At the center of U.S. efforts is compatibility with the Fair Labor Standards Act (FLSA) and other state wage-and-hour laws. The FLSA, specifically, requires employers to pay all nonexempt employees at least the minimum wage for all hours worked. If that employee works more than 40 hours in a given work week, the FLSA requires the employer to pay an overtime rate of one-and-a-half times his or her regular hourly rate for each additional hour worked. This overtime requirement is due to the employee regardless of whether the employer asked or required that the employee perform such work. If the employer had actual or constructive knowledge that the employee performed work that is an "integral and indispensable" part of the employee's principal job functions after regular hours, it must compensate the employee for that time.

While the rising tide of FLSA litigation does not yet include a significant number of electronic overtime claims, the flood appears to be coming. In fact, companies such as Verizon, T-Mobile, and Black & Decker already have been sued for unpaid overtime related to smartphone use. In one high-profile case, a police sergeant sued the City of Chicago for lost wages under the terms of the FLSA. The officer, Sgt. Jeffrey Allen, claims that over a three-year period, the police department required him and others in his division to check their BlackBerrys while off duty. In some cases, this was several hundred times per month. Allen alleges that the city never counted this electronic "checking in" as "hours worked" or paid overtime pay for it, even though it exceeded the 40 hours-per-week legal threshold. Since filing the original claim, Allen has been joined by more than 50 other officers in this class action suit.

In response to a growing number of similar FLSA cases, the U.S. Department of Labor Wage and Hour Division announced in early 2015 a request for information (RFI) regarding the use of technology by nonexempt employees to engage in off hours work to determine if any revisions to the legislation are needed. Additionally, the Department of Labor recently updated its overtime regulations, most notably, to raise the minimum weekly salary threshold to be considered "exempt" under the FLSA. Both of these efforts could force employers to take much more proactive and explicit steps to addressing the issue of off-hours work for both nonexempt and exempt employees. In particular, under the former overtime regulations, most workers are ineligible for overtime once their salary exceeds $23,660 a year. Under the new regulations, the overtime threshold at approximately $47,500 annually for a full-year worker, which increased the number of workers eligible for overtime by about 5 million and essentially force many organizations to ban off-hours work (i.e., TASW) or risk a significant increase in labor costs.

In advance of these changes in labor policy in labor policy, many employers have begun to implement these restrictions by investing in programs like Enforced Vacation that restrict emails being sent on weekends or off hours. Noguchi (2015) described such an initiative at Vynamic, a small healthcare consulting company. In response to growing concerns over off-hours work among nonexempt employees, the firm now shuts email down on weekends, and between 10:00 p.m. and 6:00 a.m. on weekdays. In addition to avoiding overtime pay and strengthening the boundaries between work and home, the new policy has had the unintended effect of making employees more aware of the need for psychologically distancing themselves (or detaching) from work as well. As an example, managers cite a new language that has developed around the policy, including terms like "zbombs" that reference substantive, thought-provoking emails sent at the end of the day that keep employees *thinking about* work off hours.

While U.S. policymakers appear focused on compensation-related issues, this last point illustrating the psychological stress associated with off-hours connectivity and the inability to detach prompted German Labor Minister Andrea Nahles to call for "anti-stress regulation" compelling companies to reduce stress in the workplace. Such regulations would ban employers from contacting employees after hours just as current policy restricts employers from contacting employees while on vacation. Some notable companies, including Volkswagen and BMW, already adopt the no-contact policy. However, it is not clear that the legislation will gain traction as Chancellor Angela Merkel has pushed back against the proposal. Not only are issues of organizational productivity at play, but so are concerns about how to monitor off-hours behavior in a reasonable, noninvasive way.

Collectively, this movement at the national level—both in the U.S. and abroad—highlights the complexities involved in managing technology use and balancing employee desires for greater flexibility and control over their demands with concerns over pay, overwork, stress, and health. The issue is complex and the solutions are nebulous at best. In the coming years, we expect that the court system, lawmakers, and policy advocates will continue to grapple with these complexities and the role of government in shaping law and influencing employment practice.

Directions for Future Research

It is clear that advances in technology have made our world a better place. In the workplace, technology provides employees with greater flexibility, improves productivity, and makes work more interesting. Yet we offer a cautionary tale about jumping too quickly to embrace the full capabilities of MCT-enabled work as it can simultaneously impede performance by creating coordination challenges and unnecessary distractions. More alarmingly, reliance on MCTs to work anywhere, anytime may put employees at risk for greater stress and health-related issues.

In this chapter we have attempted to provide a balanced perspective on both the benefits and drawbacks of technology and to highlight the organizational and national policy implications of these outcomes. Our hope is that in doing so, we may encourage more research and more critical discussions among practitioners about how employers might engage in what we term, paradoxically, *regulated flexibility*: those practices and new modes of working that leverage the benefits afforded by MCTs while also mitigating the potential for distraction, overwork, burnout, and stress. As new policies and practices begin to emerge at both the organizational and societal level, more research and evaluation efforts are needed to assess their impact and to identify those policies and practices that protect the interests of employees, organizations, and the broader community.

Limitations of the Study

In this chapter, we have drawn upon our own research data as well as findings from the broader scholarly and practitioner literatures to identify the paradoxes associated with MCT use and the policies and practices that might govern such use. While our own data include robust samples totaling more than 500 employees from a range of industries and occupations, there are limitations to our work, and the work we have cited here, that should be acknowledged. Most notably, the data collected by our research team to date are entirely from U.S.-based samples. While the broader literature is not restricted in the same way, it is the case that most of the work we cite is largely conducted in Western contexts. Accordingly, our interpretations of the data and recommendations for policy and practice do not fully account for cultural differences in other parts of the world. For instance, the limitation of work hours may be more socially acceptable in countries with a lower performance orientation (i.e., those who place less value on performance and high standards), including Nordic countries such as Sweden (House, Hanges, Javidan, Dorfman, & Gupta, 2004). However, such a practice may be viewed as counterproductive or untenable in countries that are more performance oriented (e.g., Germany, Taiwan, South Korea). Future research should explore the role that culture plays in shaping the utilization of MCTs as well as the efficacy of various practices and policies that govern their use.

ACKNOWLEDGMENT

This study is funded by a grant from the SHRM Foundation. However, the interpretations, conclusions, and recommendations are those of the authors and do not necessarily represent the views of the SHRM Foundation. The authors would like to thank other members of our research team, including Suzanne Masterson, Elaine Hollensbe, Therese Sprinkle, and Sung Doo Kim with their assistance in data collection and analysis. We also wish to thank Jeremy Hoffman and Alexandra Myers for their assistance on early versions of this chapter.

REFERENCES

American Psychological Association (APA). (2013). *Communication technology: Implications for work and well-being.* Retrieved from http://www.apaexcellence.org/assets/general/2013-work-and-communication-technology-survey-final.pdf

Barber, L. K., & Santuzzi, A. M. (2015). Please respond ASAP: Workplace telepressure and employee recovery. *Journal of Occupational Health Psychology, 20*(2), 172–189. doi: 10.1037/a0038278

Blair-Loy, M. (2009). Work without end? Scheduling flexibility and work-to-family conflict among stockbrokers. *Work and Occupations. 36*(4), 279–317. doi: 10.1177/0730888409343912

Brewer, P. E. (2015). *International virtual teams: Engineering global success.* Piscataway, NJ: Wiley.

Deci, E. L., & Ryan, R. M. (2012). Self-determination theory. In P. A. M. Van Lange, A. W. Kruglanski, & E. T. Higgins (Eds.), *Handbook of theories of social psychology* (Vol. 1, pp. 416–433). Thousand Oaks, CA: Sage.

Empirica. (2000). *Electronic commerce and telework trends: Benchmarking progress on new ways of working and new forms of business across Europe.* Retrieved from http://www.ecatt.com/freport/ECaTT-Final-Report.pdf

Fenner, G. H., & Renn, R. W. (2010). Technology-assisted supplemental work and work-to-family conflict: The role of instrumentality beliefs, organizational expectations and time management. *Human Relations, 63*(1), 63–82. doi: 10.1177/0018726709351064

Furst-Holloway, S., & Jones, T. M. (2011). *HR's role in managing the virtual workplace.* Prepared for the Greater Cincinnati Human Resource Association.

Galinsky, E., Bond, J. T., & Hill, E. J. (2004). *When work works: A status report on workplace flexibility: Who has it? Who wants it? What difference does it make?* Retrieved from http://familiesandwork.org/3w/research/downloads/status.pdf

Good. (2012). *Good Technology survey reveals Americans are working more, but on their own schedule.* Retrieved from https://www1.good.com/about/press-releases/161009045.html.

Haeger, D. L., & Lingham, T. (2014). A trend toward work-life fusion: A multigenerational shift in technology use at work. *Technological Forecasting & Social Change, 89,* 316–325. doi: 10.1016/j.techfore.2014.08.009

Hampton, K. N., Sessions, L. F., Her, E. J., & Rainie, L. (2009). *Social isolation and new technology.* Washington, DC: Pew Internet & American Life Project. Retrieved from http://www.pewinternet.org/Reports/2009/18–Social-Isolation-and-New-Technology.aspx

Harris, K. J., Marett, K., & Harris, R. B. (2011). Technology-related pressure and work–family conflict: Main effects and an examination of moderating variables. *Journal of Applied Social Psychology, 41*(9), 2077–2103. doi: 10.1111/j.1559-1816.2011.00805.x

Henle, C. A., & Blanchard, A. L. (2008). The interaction of work stressors and organizational sanctions on cyberloafing. *Journal of Managerial Issues, 20*(3), 383–400.

Henle, C.A., Kohut, G., & Booth, R. (2009). Designing electronic use policies to enhance employee perceptions of fairness and to reduce cyberloafing: An empirical test of justice theory. *Computers in Human Behavior, 25*(4), 902–910. doi: 10.1016/j.chb.2009.03.005

House, R. J., Hanges, P. J., Javidan, M., Dorfman, P. W., & Gupta, V. (Eds.). (2004). *Culture, leadership, and organizations: The GLOBE study of 62 societies.* Thousand Oaks, CA: Sage.

International Labor Office (ILO). (2004). *Economic security for a better world.* Geneva, SUI: Author.

Jarvenpaa, S., & Lang, K. (2005). Managing the paradoxes of mobile technology. *Information Systems Management, 22*(4), 7–23. doi: 10.1201/1078.10580530/45520.22.4.20050901/90026.2

Jarvenpaa S. L., Lang K. R., & Tuunainen, V. K. (2005). Friend or foe? The ambivalent relationship between mobile technology and its users. In C. Sørensen & Y. Yoo (Eds.), *Designing ubiquitous information environments: Socio-technical issues and challenges* (pp. 29–42). New York, NY: Springer.

Kelly, E. L., Moen, P., & Tranby, E. (2011). Changing workplaces to reduce work–family conflict: Schedule control in a white-collar organization. *American Sociological Review, 76*(2), 265–290. doi:org/10.1177/0003122411400056

Kim, S., Bologna, D., Furst-Holloway, S., Hollensbe, E. C., Masterson, S. S., & Sprinkle, T. A. (2014). Taking a break via technology? Triggers, nature, and effects of online work breaks. *Proceedings of the National Meeting of the Academy of Management,* Philadelphia, PA.

Kossek, E. E., Thompson, R. J., & Lautsch, B. A. (2015). Balanced workplace flexibility: Avoiding the traps. *California Management Review, 57*(4), 5–25. doi: 10.1525/cmr.2015.57.4.5

Lanaj, K., Johnson, R. E., & Barnes, C. M. (2014). Beginning the workday yet already depleted? Consequences of late-night smartphone use and sleep. *Organizational Behavior and Human Decision Processes, 124*(1), 11–23. doi: 10.1016/j.obhdp.2014.01.001

Lautsch, B. A., Kossek, E. E., & Eaton, S. C. (2009). Supervisory approaches and paradoxes in managing telecommuting implementation. *Human Relations, 62*(6), 795–827. doi: 10.1177/0018726709104543

Leung, L. (2011). Effects of ICT connectedness, permeability, flexibility, and negative spillovers on burnout and job and family satisfaction. *Human Technology, 7*(3), 250–267. doi: 10.17011/ht/urn.2011112211714

Lim, V. K. (2002). The IT way of loafing on the job: Cyberloafing, neutralizing and organizational justice. *Journal of Organizational Behavior, 23*(5), 675–694. doi: 10.1002/job.161

McPherson, M., Smith-Lovin, L., & Brashears, M. E. (2006). Social isolation in America: Changes in core discussion networks over two decades. *American Sociological Review, 71*(3), 353–375. doi: 10.1177/000312240807300610

Nafukho, F. M., Graham, C. M., & Muyia, H. M. A. (2009). Harnessing and optimal utilization of human capital in virtual workplace environments. *Advances in Developing Human Resources, 12*(6), 648–664. doi: 10.1177/1523422310394791

Noguchi, Y. (2015). *Amid new overtime rules, more employers might set email curfew.* Retrieved from http://www.npr.org/sections/alltechconsidered/2015/07/08/416515537/amid-new-overtime-rules-more-employers-might-set-email-curfew

Perlow, L. (1999). The time famine: Toward a sociology of work time. *Administrative Science Quarterly, 44*(1), 57–81. doi: 10.2307/2667031

Perlow, L. A., & Porter, J. L. (2009). Making time off predictable—and required. *Harvard Business Review, 87*(10), 102–109. doi: 10.1108/hrmid.2010.04418bad.012

Ran, B. (Ed.). (2013). *The dark side of technological innovation.* Charlotte, NC: Information Age.

Richardson, K., & Benbunan-Fich, R. (2011). Examining the antecedents of work connectivity behavior during non-work time. *Information and Organization, 21*(3), 142–160. doi: 10.1016/j.infoandorg.2011.06.002

Rouse, C. (2014). The economics of workplace flexibility. Retrieved from https://www.whitehouse.gov/blog/2010/03/31/economics-workplace-flexibility

Sarker, S., Xiao, X., Sarker, S., & Ahuja, M. (2012). Managing employees' use of mobile technologies to minimize work–life balance impacts. *MIS Quarterly Executive, 11*(4), 143–157.

Science Alert. (2015). *Sweden is shifting toward a 6-hour work day.* Retrieved from http://www.sciencealert.com/sweden-is-shifting-to-a-6-hour-workday

Shortland, S. (2005, Summer). *Work–life balance: The impact of policy and practice on the human resource function.* Retrieved from http://www.ircounselors.org/irconcepts/2005summer.html

Stangler, C. (2015): *After-hours use of work email may finally see some labor regulations.* Retrieved from http://www.ibtimes.com/after-hours-use-work-email-may-finally-see-some-labor-regulations-1940882

Turel, O., Serenko, A., & Bontis, N. (2011). Family and work-related consequences of addiction to organizational pervasive technologies. *Information & Management, 48*(2), 88–95. doi: 10.1016/j.im.2011.01.004

Van Dyne, L., Kossek, E. E., & Lobel, S. (2007). Less need to be there: Cross level effects of work practices that support work–life flexibility and enhance group processes and group-level OCB. *Human Relations, 60*(8), 1123–1154. doi: 10.1177/0018726707081657

Vitak, J., Crouse, J., & LaRose, R. (2011). Personal internet use at work: Understanding cyberslacking. *Computers in Human Behavior, 27*(5), 1751–1759. doi: 10.1016/j.chb.2011.03.002

WorldatWork. (2009). *Telework trendlines 2009: A survey brief from WorldatWork.* Retrieved from http://www.worldatwork.org/waw/adimLink?id=31115

CHAPTER 6

JAPAN'S SCIENCE AND TECHNOLOGY INNOVATION NATION STRATEGY AND THE PATH OF ECONOMIC GROWTH

With a Look at China's Innovation-Driven Strategy

Tong Quan
Shanxi University, China

ABSTRACT

This chapter examines the circumstances surrounding the Japanese government's introduction of the "Science and Technology Innovation Nation Strategy" in 1995, focusing on issues such as the relationship between investment input into technology innovation and economic growth, compares this with China's current policy framework on innovation capacity, and puts forward some recommendations at the policy and institutional levels for technological innovation for China to strengthen future innovation endeavors.

Technological Innovation Networks, pages 123–136

Led by high volumes of investment, and after experiencing very rapid growth in the better part of the past 20 years, the Chinese economy is currently facing new dilemmas. Firstly, the marginal benefits of the investment-led approach are tapering off; the high-consumption plus high-GDP model of growth has resulted in a multitude of developmental problems such as resources misuse and environmental pollution. Secondly, as higher education has become more widely available, people's income has risen as labor cost is also continuing to rise; China's status as the global factory is steadily declining each year, and the country's ability to attract capital is starting to weaken. The investment-led model of growth can no longer remain the fundamental force to stimulate economic growth. In November 2011, the CPC convened the Third Plenary Session of its 18th Central Committee of the Chinese Communist Party of the National Congress; this assembly formally approved the "innovation-driven" strategic policy decision to promote sustainable and stable economic development by designating independent R&D to achieve technological innovation as the main engine for driving economic growth. Since the end of WWII, Japan's national development strategy has gradually transitioned from "trading nation," to "technology nation" to "science and technology innovation new strategy," and the country has moved further in the areas of information systems and intellectual property protection. Even though Japan formulated its economic development strategies on the basis of detailed analyses and research studies on the domestic and international economic development trends, judging by the end results, economic growth has not moved toward the expected objectives. As China has put forward its innovation-driven national strategy, would this be sufficient to ensure painless attainment of the target objectives of doubling GDP output and doubling average national income by 2020 as put forward by the 18th CPC National Congress?

Although different countries have different systems of government and the degrees of state intervention in the economy differ, under the greater backdrop of economic development in the same global environment, national economic development strategy concerns a country's status and influence within the global economy. For this reason, taking stock of Japan's economic development outcomes stemming from the introduction of the science and technology innovation strategy while analyzing the root causes of the problems would be conducive for implementing China's innovation-driven strategy.

THE POLICY BASIS FOR JAPAN'S NATIONAL INNOVATION SYSTEM

From the end of WWII to this day, the Japanese government has focused attention on economic development with the introduction of a series of different economic development strategies in different periods. The country roughly experienced the five phases of building the nation by trade, building the nation by technology, building the nation by information technology, building the nation by intellectual property, and building the nation by science and technology innovation.

During the period immediately after WWII, Japanese society was acutely impoverished, depleting 25% of its national wealth with extreme material scarcity (Hyohe, 1985, p. 336) and a steep drop in industrial and agricultural outputs. Adding to the hardship were coexisting hyperinflation, disarrayed consumption, derelict agricultural lands, and gradually dwindling food supplies from Taiwan and Southeast Asia, thus further exacerbating the food crisis. Japan experienced a 40-year record bad harvest in 1945, as food production was cut by 33.2% compared to the previous year, with total harvest amounting to merely 60% of average yield. In each average day, 2.5 pedestrians would be fainting on the street in the Tokyo-Ueno areas due to malnutrition; in addition, there was a stream of news reports of people dying from starvation. In view of situation at the time (Sadakichi, 1996, p. 2), the Japanese government adopted the "trading nation" strategy, and beginning from 1955, Japan's economy heralded an era of rapid growth, recording a 40% increase in 1956 compared with 1955; in the 3 years from 1959 to 1961, the Japanese economy maintained growth rates of 9.3%, 13.1%, and 11.6% (Ryoko, 1989, p. 288). By the 1960s (Sadakichi, 1996, p. 83), the Japanese Prime Minister at the time Ikeda Hayato put forward a "national income doubling plan" with the aim of increasing national income by twofold within a decade; this objective was in fact achieved in less than 10 years. However (Li, 2002), due to its scarcity in resources endowment, Japan's advantages engendered from the trading nation strategy were blown away by the impacts of the Oil Shock. Nevertheless, the accumulation of national wealth laid down a solid economic foundation for subsequent national strategic adjustments.

In 1975, in a paper titled "Transition from a Processing and Trading Nation to the Technology Nation," Professor Saitou of the Faculty of Economics of Chuo University proposed for the first time the "technology nation theory." This puts scientific and technological progress as the bedrock of national development in economic, political, cultural, and other spheres. In 1980, the Ministry of International Trade and Industry (MITI) formally proposed the technology nation strategy in *MITI Policy Perspectives for the '80s* (Ministry of Education, Science and Culture, 1980); 1981 saw the Japanese government introducing the two regulatory provisions of "promoting innovative science and technology institutions" and "institutional systems for research into the next generation of industrial infrastructure technology," adopting a series of incentive measures and preferential policies to kick-start scientific and technology development by affirming electronics technology, IT, biosciences, transportation technology, and energy technology as the focuses of scientific and technological development in the 1980s.

In 1987, British economist C. Freeman proposed the notion of a "national system of innovation." The interrelationships between various entities in the public sector and the private sector formed a network system that is conducive for promoting technology generation, introduction, and diffusion (Freeman, 1989, p. 34). He also conducted detailed studies of the roles of MITI and industry R&D, especially R&D relevant to technology introduction, the roles of education and training and relevant social innovation, and the structures of corporate groups. In the advent of Freeman's system of national innovation, countries gradually turned their attention to building up their own systems of national innovation, as Japan was starting to emphasize the role played by the state in technology innovation, at a time when the country's economy was under double pressure from domestic and international fronts. In 1995, the Japanese legislature passed the Science and Technology Basic Law, which explicitly stipulated the strategy of "science and technology innovation nation" as a basic national policy; the three phases of detailed implementation measures for the science and technology innovation nation strategy were formulated in 1996, 2001, and 2005 respectively. "Information technology nation" and "intellectual property nation" that appeared thereafter may be regarded as important steps of the strategy of "science and technology innovation nation."

On the basis of the strategy of building the nation by science and technology innovation and as the Japanese economy was remaining in hiatus through the 1990s, Japanese scholar Hiroshi Takahashi conducted research into Japan's IT revolution innovation policies from 1950 to 2000 and then proposed that the Japanese government had not been able to foster a conducive environment for sustainable development, as Japanese cyber space was closed off to the outside world; the government and the market were interventionist in nature that only provided technological development support to a selected few suppliers. Therefore, in order for Japan to extricate itself from economic recession, the country must clearly define new pillar industries for itself. On this account, the Japanese government formulated the national economic development strategy of "building a highly advanced information society" with the IT and communication industries as leaders to drive industry transformation and promoting economic development. In November 2000, the Japanese government formally introduced the "basic national IT strategy," at the same time designating a ministerial post for management of the IT sector with the expectation of realizing full automation in the domestic economy as soon as possible.

Nevertheless, the IT nation strategy did not turn around the economic depression started since the 1990s, and at a time when the international competitiveness of Japanese industries was slipping the Japanese government focused attention into formulating the "intellectual property" strategy by leveraging prime ministerial policy decision mechanisms. In 2002, the Japanese government convened four sessions of intellectual property strategy meetings and released the "Outlines of Japan Intellectual Property Strategy," which clearly elucidated the basic direction for the "intellectual property nation" strategy, including a detailed action plan covering the period up to 2005.

In the first place, the universities, publicly funded research bodies, and business enterprises formed into a core with attention to application of technological innovation in industry while promoting a constant stream of intellectual and invention innovations, refining the intellectual innovation institutional systems; strengthening the assessment and evaluation of scientific research results; formulating rules that can stimulate scientific and technological innovation inventions; and promoting collaborative research between industry, the government, and academic institutions. Secondly, the focus was on facilitating intellectual property transfer, developing system software for trade and exchange of intellectual properties and accelerating turnover and utilization of intermediate and semi-finished products, implementing practical strategies for commercialization of invention innovation and R&D results while maximizing the socioeconomic benefits. Thirdly, the strategy was to protect and foster new product varieties and regional brands by legislating to hit at distribution of imitation and counterfeit products, establishing specialized higher courts to speed up the hearing process for patent-related cases, and setting up IP headquarters at universities and so on. Fourthly, training outstanding human talents with rich professional knowledge was made a priority.

In 2003, in order to further improve the international competitiveness of Japanese industries and strengthen intellectual property development and protection, the cabinet office established a dedicated general office for IP strategy to continuously update the science and technology development strategies for the different sectors.

SCIENCE AND TECHNOLOGY INVESTMENT DURING IMPLEMENTATION OF THE SCIENCE AND TECHNOLOGY INNOVATION NATION STRATEGY AND THE PATH OF ECONOMIC GROWTH

After Japan affirmed the Science and Technology Innovation Nation Strategy, the country formulated three "national science and technology basic plans" in 1996, 2001, and 2005, each lasting a period of five years, while the government, industry, the higher education institutions along with other research bodies respectively contributed 17.6, 18.8, and 25 trillion yen, focusing on support for such fields as life science, information science, environmental science, and new nano-materials. Starting from the end of the second phase, the Ministry of Economy, Trade, and Industry (METI) introduced a new industry innovation strategy by affirming the designation of fuel cell, smart information appliances, robotics, creative industry, health and welfare industries, environmental and energy industries, and corporate business services as pillar industries supporting future economic development in Japan and thus given significant government support. In 2005 (Bunro, 2006), investment outlay into life science, information science, environmental science, and new nano-materials respectively increased by 11.1%, 18.1%, 63.0%, and 14.4% compared to 2001, but investment outlay into the areas of energy, manufacturing technology, social infrastructures, and cutting-edge technology declined respectively by 2.9%, 12.7%, 5.3%, and 40.2%, respectively; however, these still accounted for the lion's share of funding allocation (Bunro, 2006).

Japan's total R&D expenditure was 17.3791 trillion yen during the 2011 financial year, including industry R&D expenditure of 12.2718 trillion yen, accounting for 70.6% of total expenditure, while R&D expenditures by government and nonprofit organizations were 1.5668 trillion yen, accounting for 9.0% of total expenditure; R&D expenditures by higher education institutions were 3.5405 trillion yen, accounting for 20.4% of total expenditure. R&D expenditure by industry declined slightly to 12.1705 trillion yen during the 2012 financial year, while the figures for government, nonprofit organizations, and higher education institutions recorded some increase at 1.5917 trillion yen and 3.5624 trillion yen respectively, giving a total of 17.3426 trillion yen, declining slightly over 2011 (Ministry of Internal Affairs and Communications, 2013). For industry, science and technology R&D expenditure by the manufacturing industry accounted for the highest proportion in 2012, at 59.7% of total expenditure to top 7.2657885 trillion yen. This figure included the transport machinery at 18.7%, information and communication apparatus manufacturing at 13.7%, electrical machinery and appliances manufacturing at 8.4%, business use mechanical equipment manufacturing at 7.6%, welding machine manufacturing at 2.3%, production machinery at 3.6%, with electronics parts and components and IC manufacturing at 5.4% of total expenditure. The pharmaceutical manufacturing industry came closely behind at 10.7% of total expenditure (Ministry of Internal Affairs and Communications, 2013). (Refer to Figure 6.1.)

Looking at the number of personnel in research, in 2012, Japan had approximately 646,000 personnel engaged in research into science and technology innovation; the United States had up to 1,413,000 scientific research personnel, while China has 1,318,000 personnel. Japan's absolute number of research personnel is less than half that of the United States or China. However, when considering the number of research personnel as

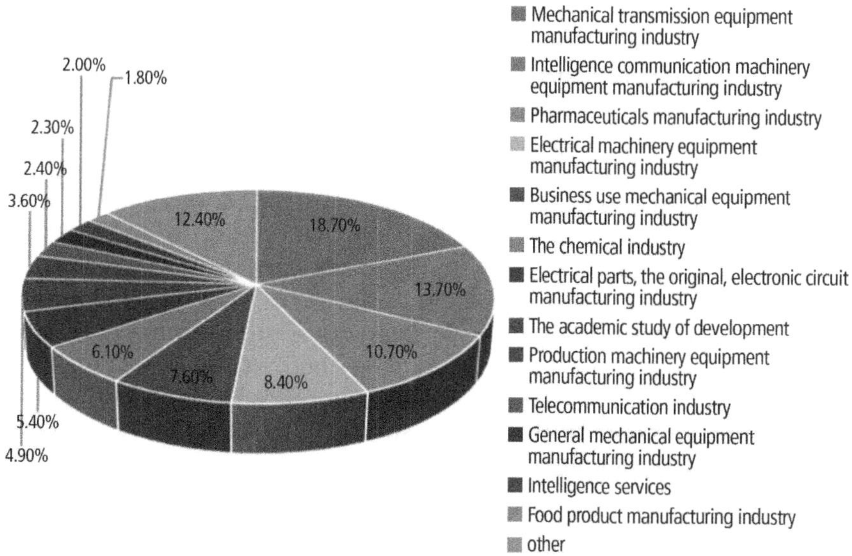

Figure 6.1 2012 science and technology research expenditures by Japanese industries as proportions of total industry expenditure (Ministry of Internal Affairs and Communications, 2013).

a proportion of total population, Japan has 50.8 per ten thousand, a little less than the figure for South Korea of 58.0 per ten thousand, the United States has 46.8 per ten thousand, and China has only 9.8 per ten thousand. In terms of research expenditure per head of population, Japan has been experiencing a downward trend for research expenditure per head of population since 2009, falling from 22.41 million yen to 20.73 million yen in 2013. Japan's research expenditure per head of population of USD253,123 in 2012 was close to the U.S. level of $269,062 in 2007, and China's 2011 figure was just reaching USD157,935 (Ministry of Internal Affairs and Communications, 2013).

The science and technology nation strategy was proposed under a backdrop of the appearance of zero growth or negative growth after bursting of the bubble economy in Japan; it was a measure by the Japanese government to turn around weakening economic growth by attempting to leverage the country's own technological innovation capacity. However, the country's economic growth has not seen much sign of a pickup since the 1990s. In that case, have the investment expenditures into science and technology innovation had any expected effect in promoting economic growth in Japan? We will use the data compiled by the Statistics Bureau of the Ministry of Internal Affairs and Communications on science and technology investment outlays over the years (research expenditures), total GDP, and so on to undertake a study by applying quantitative methodology.

From a theoretical standpoint, applying technology innovation expenditure to accelerate economic growth has two lagging effects—that is, the lags in input and output. The first is the input stage during research and development: The lagging effect in technology innovation expenditure involves a process of "technology innovation input—accumulation of relevant knowledge—success of technology innovation." In other words, a certain amount time is needed from outlaying investment funds to success in research and development. Secondly, there is the promotion and application phase after output of the innovation product; the output lag involves a process of "success in technology innovation—patent application—trial utilization—promotion—capturing market." In other words, a certain amount of time is needed from success in technology innovation to full-scale market promotion.

Assuming the chosen developer to outlay investment for technology innovation already has an innovative team with relatively rich relevant knowledge, which can diminish the lagging problem of technology innovation, however, the output lagging effect will still be an unavoidable issue. Existing research indicated R&D expenditure outlay would normally only show effect after more than two years (He, 2003). Therefore, we will exclude the current period R&D expenditure to examine the lag in the boost to Japan's GDP.

In order to explore the boosting effects on nominal GDP (unit: JPY 000m) from R&D expenditure, a unit root test of the 1995–2013 R&D expenditure on GDP is carried out to obtain the two sets of figures that both meet the criteria for first-order difference smoothness; next, the software package Eviews 7.0 is used to derive the correlation coefficient $r = 0.70$, which indicates the first-order difference series on R&D expenditure and nominal GDP has a relatively high degree of positive correlation (Table 6.1).

The expenditure method can also be used to show that R&D expenditure input would have a certain degree of positive boosting effect on the current year nominal GDP, coupled with the linear relationship in its first-order difference. The iterative method can be applied to derive the presence of a negative correlation between nominal GDP and the lagged years' R&D expenditure input; moreover, the 4-year lagged R&D expenditure input has the highest negative correlation (Table 6.2). The software package

TABLE 6.1 A Unit Root Test of the 1995–2013 R&D Expenditure on GDP

	D (R&D exp.)	D (nom. GDP)
D (R&D exp.)	1.000000	0.697491
D (nom.GDP)	0.697491	1.000000

TABLE 6.2 A Negative Correlation Between Nominal GDP and the Lagged Years' R&D

	R&D exp. (–4)	Nom. GDP
R&D exp. (–4)	1.000000	–0.804236
Nom. GDP	–0.804236	1.000000

Eviews 7.0 can also be used to derive the correlation coefficient of $r = -0.80$ to verify this.

Summing up the above, R&D expenditure input has a positive effect on the current year nominal GDP, and the science innovation result stemming from R&D input has no lag effect, or its lag effect is negative. This may also be demonstrated by the variations in R&D input and in the proportion of GDP.

Figure 6.2 is Japan's technology innovation investment input and its variation as a proportion of GDP, which shows a correlated relationship; that is, higher science and technology R&D input means slightly higher current year nominal GDP, and, conversely, lower science and technology R&D input means slightly lower current year nominal GDP, and the magnitudes are quite similar. This is also another way of saying scientific research input has a direct effect on GDP. However, without considering other factors affecting GDP, with the nominal method of calculating GDP, science and technology innovation input would have to be added to total GDP, then after taking into consideration the lagging effect of technological innovation, what is reflected by such a positive correlation is merely the GDP boost arising from R&D input, rather than the GDP boosting effect of the research results being transformed into productivity. Hence, the conclusion can be

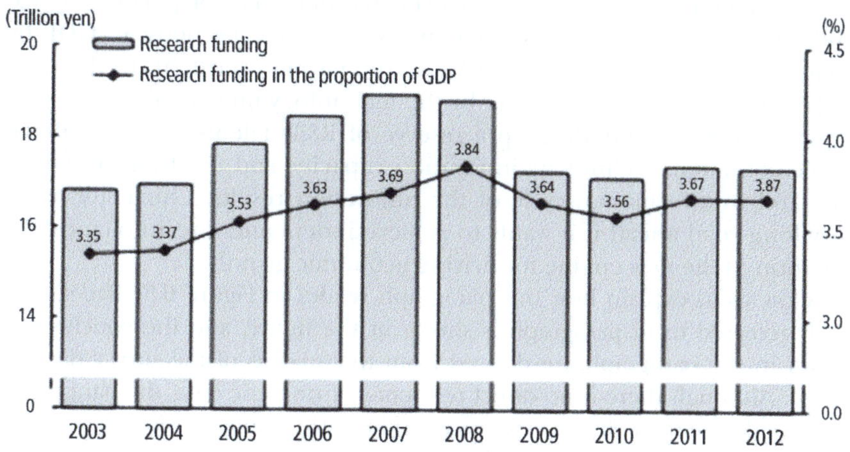

Figure 6.2 R&D input as a proportion of GDP for the period 2003–2012.

drawn that, apart from the investment effect, contribution to Japan's economic growth from science and technology innovation inputs during the period 2003–2012 was 0, or it may even had a negative effect.

THE IMPLEMENTATION BASIS AND OUTLOOKS FOR CHINA'S INNOVATION-DRIVEN STRATEGY

Similar to Japan, China's economic growth also experienced an export-dependent and investment expansion phase. Japan started the technology nation strategy in 1981 and provided funding and human resources supports for the science and technology innovation nation strategy, and subsequent implementation of the IT and intellectual property nation strategy laid the information and legislative foundations for the science and technology innovation strategy. China proposed a "science and education nation" strategy in 1995, with expanded investment input into the higher education institutions and scientific research institutes while providing funding, knowledge resources, and HR reserves for scientific and technological progress. As the global economic climate has changed, China's investment-led economic development model is no longer adequate to meet the requirements of productivity growth; the state thus explicitly defined "innovation-driven" as the long-term national development strategy in 2013.

The innovation-driven strategy is a far-sighted national economic development strategy directed at the present-day reality of low-level, high-speed economic growth in China, with the aim of raising the quality of economic growth, realizing highly efficient resources utilization, enhancing added value for Chinese products, and changing the GDP structure. However, the implementation basis and performance outcomes of Japan's science and technology innovation strategy has shown that the boost to GDP by technological innovation will not be accomplished overnight. In the light of Japan's situation, and judging by the technology innovation investment input needed and building up a reserve of R&D talents before strategic implementation, technology innovation capacity, and the transformation rate and transformation cycle of the innovation results, China still has a very long road ahead if it wants to achieve independent technological innovation as the new engine for driving economic growth.

Allow us to explain how this paragraph relates to Figure 6.3. The survey study referred to in paragraph is shown in the figure, and the conclusions drawn in the paragraph are derived from the analysis of the survey data in figure. Although there is no direct reference of numeric data, the paragraph did take the trend of data as reference. And we also emphasize the importance of the trend rather than the value itself in the name of Figure 6.3. Therefore, we took this arrangement. The results of a survey study by the Ministry

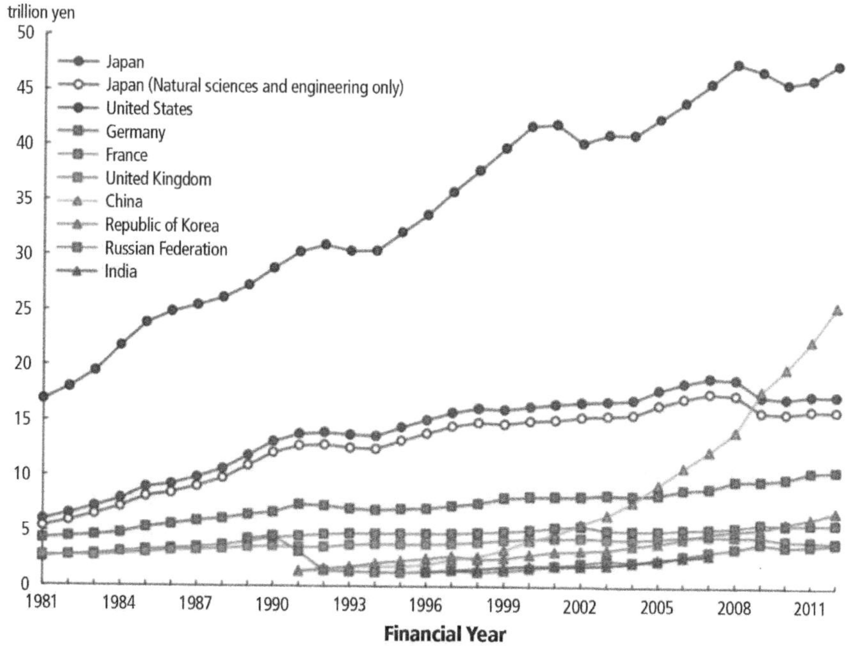

Figure 6.3 Trends in R&D expenditures in selected countries (OECD purchasing power parity conversion) (Ministry of Education, Culture, Sports, Science, and Technology, 2014).

of Education, Science and Culture of Japan indicate China started collecting statistics on investment input into technological innovation in 1991; the magnitude of investment was small at the time, in fact, smaller than that of South Korea in the earlier years. Thereafter China's investment input steadily increased year by year, reaching a level in 2009 roughly at par with Japan's investment input into technology innovation while continuing to maintain a tendency of very rapid growth. However, China's overall economic structure is highly complex: The country's population and total GDP are immense, although investment input into technological innovation is growing rapidly, but because the country's industry development strategies are not sufficiently clearly defined, the net amount of such technological innovation investment input allocated to the individual sector is far lower than Japan.

Secondly, the transformation rate of technological innovation needs to be improved. Higher education institutions usually have large groups of scientific research personnel with excellent research infrastructures, hence, they can become a country's important bases for scientific research and innovation, and they constitute one of the main sources of motivating force to promote economic growth and raising the level of national core competitiveness.

However, an existing research shows that "our higher education institutions' transformation rate of scientific research results is less than 10% and the rate of true commercialization is just around 5%, far lower than the transformation rate of developed countries" (J & L, 2013, pp. 201–204).

In the example of the higher education institutions, these bodies tend to have rigid management systems, eager for instant success, obsessed with benchmarks and indicators in their research activities and teaching assignments, and obsessed with bureaucratic job title appraisal. The piece-work system of performance assessment is seriously hampering innovative thinking by the scientific and technological researchers, which also means these higher education institutions are unable tolerate those researchers who have ideas but cannot quickly produce scientific research results within a given period of time. Under the dominance of such an institutional establishment, innovative research projects are shelved by researchers because these have long cycles and high costs; supplanted in their place are academic results that can meet publication deadlines in the academic journals. The academic journals almost invariably become the dominant factor in academic research, as the journal requirements come far ahead of the appeals of social requirements to the academic scholars. This situation will most certainly lead to decoupling of the research field and market demand, thus adversely affecting the rate of transformation of scientific research results and further impeding the boosting effects of technological innovation on the national economy.

A more rational scheme in channeling investment into science and technology innovation by the business sector, government, and higher education institutions would be capable of effectively promoting progress in science and technology. As indicated in Figure 6.4, it can be readily seen from the makeup of investment input into technological innovation in the three countries of U.S., Japan, and Germany that industry input is firmly leading the field with a stable growth rate, while investment inputs by government and higher education institutions are holding steady, or perhaps government investment input is slightly more than that of the higher education institutions. Industry input started to overtake government input in China after 1999, reaching a level of more than twice the combined government and higher education inputs by 2012. The rapid increase in industry investment input into science and technology is reflection of a constantly expanding industry needs for technological innovation; however, the rate of transformation of technological innovation remains at low level, with the U.S., Germany, and Japan scoring approximately 40%, while the figure for China is only 10% (Financial Channel of Chinanews, 2013). Putting aside the question of market prospects after transformation of the science and technology results and the boosting effects to GDP, by just focusing on investment returns in isolation, it is not entirely appropriate to mention China's investment returns on its innovation investment input side by side

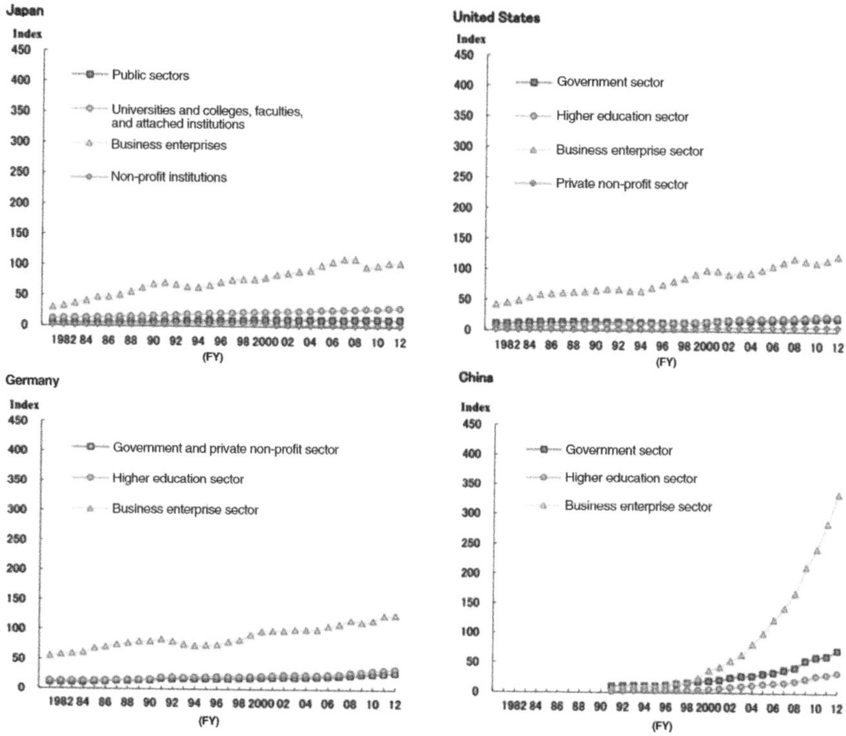

Figure 6.4 Key structural features of R&D investment in the four countries of Japan, U.S., Germany, and China (Ministry of Education, Culture, Sports, Science, and Technology, 2014).

with Japan's negative correlation effect. Moreover, the projects receiving government investment input for scientific and technological innovation in this country are to a large extent an embodiment of the government's overall strategic intents of safeguarding China's international position and economic development, rather than a practical posture dictated by market demands or if these projects can truly stimulate economic growth. Nevertheless, because of the state's dominant role in scientific research, the aura and status of the research projects undertaken in higher education institutions and scientific research institutes are unmatched by the status of the projects receiving industry investment inputs.

Although the proportion of industry input into science and technology investment is far higher than government and higher education institutions in China, due to the government's leading role and the technological innovation activities influenced by the dominant position of the academic journals are detached from the market, and a general attitude of underrating the industry's scientific and technological innovation, it would be difficult for the level of technological innovation transformation and the rate of technological innovation transformation to adapt to the goal of a future innovation-driven strategy. Therefore, for China's scientific and technological innovation to transition from government dominance to a model led by the market, industry, and business enterprises is the truly effective avenue for higher education institutions to improve their rate of transformation of scientific and technological innovation. It is also central to China's goal of breaking away from the investment-led development model and fulfill the country's vision of an innovation-driven strategy.

REFERENCES

Bunro, S. (2006). *Science and technology basic plan (draft) and competitiveness research funding programs.* Retrieved from http://www.jst.go.jp/po_seminar/h17semi/pdf/02shiozawa.pdf

Freeman, C. (1989). *Technology policy and economic performance.* London, England: Pinter.

Financial Channel of Chinanews. (2013). *Transformation rate of science and technology results in China.* Retrieved from http://finance.chinanews.com/cj/2013/12-21/5647840.shtml

He, W. (2003). The output effects of R&D expenditure by medium to large industrial enterprises in this country—1990–2000 case study analysis of medium to large industrial enterprises data. *Economic Science, 3,* 5–11.

Hyohe, O. (1985). *Economic statistics collection of Japan.* Tokyo: Japan Lei Ru Shinsha.

J, K., & L, Q. (2013). A preliminary exploration into the causes of the relatively low rate of transformation of scientific research results in the higher education institutions. *Technology & Innovation Management, 3,* 201–204.

Li, P. (2002). The role of Japan's 'nation trading' strategy in capital accumulation in the post-war period. *Japan Issues Studies, IV,* 1–5.

Ministry of Education, Culture, Sports, Science, and Technology. (1980). *MITI policy perspectives for the '80s.* Retrieved from http://www.mext.go.jp/b_menu/hakusho/html/hpaa198001/hpaa198001_2_020.html.

Ministry of Education, Culture, Sports, Science, and Technology. (2013). *MEXT. Research expenditure.* Retrieved from http://www.mext.go.jp/component/b_menu/other/__icsFiles/afieldfile/2014/09/26/1351776_03.pdf

Ministry of Education, Culture, Sports, Science, and Technology. (2014). *MEXT. Research expenditure.* Retrieved from http://www.mext.go.jp/component/b_menu/other/__icsFiles/afieldfile/2014/09/26/1351776_03.pdf

Ministry of Internal Affairs and Communications. (2013). *2013 survey results on science and technology research.* Retrieved from http://www.stat.go.jp/data/kagaku/kekka/youyaku/pdf/25youyak.pdf

Ryoko, N. (1989). *A socioeconomic history of Japan.* Tokyo, Japan: Showado.

Sadakichi, K. (1996). *A review of the post-war Japanese economy.* Tokyo, Japan: Gakubunsha.

CHAPTER 7

PATENTS AND PERFORMANCE

The Case of Immigrant Entrepreneurs in the United States

Trayan Kushev
Cal Poly Pomona

John M. Mueller
California State University–Fresno

ABSTRACT

Researchers have been intrigued about the impact of immigrants on many phenomena, such as wages, taxes, and society in general (Card, 2009). Yet we know little of how immigrants in the United States perceive the patent system, a key mechanism for promoting innovation. This study addresses this gap by examining whether immigrants in the United States value the American patent system, and if so, whether the patent system helps immigrants with the performance of their new firms. Multiple statistical analyses, including *t*-tests, ANOVA, Cox proportional hazard modeling (survival analysis), and panel data regression are used on a longitudinal dataset, the Kauffman Firm Survey,[1] with six years of data. The results suggest that, when building their own businesses, immigrants use the American patent system more than people

Technological Innovation Networks, pages 137–165

born and raised in the United States, thus supporting Hunt and Gauthier-Loiselle (2010) and Peri and Sparber (2011) in the context of immigrants as founders as opposed to employees of a firm. Patents have little effect on the survival and growth of firms founded by immigrants, which is also the case with firms founded by Americans. The findings of the study are important to policymakers, with the results suggesting that the patent system is beneficial in enticing immigrants to start innovative firms in the United States rather than building the firms in their native countries.

The United States is commonly referred to as a country of immigrants. Not surprisingly, researchers have been intrigued about the impact of immigrants on many phenomena, such as wages, taxes, and society in general (Card, 2009). In an entrepreneurial sense, immigrants made up over 25% of founders of public venture-backed companies in the 15-year period between 1990 and 2005 (Anderson & Platzer, 2006) and are more likely to start companies compared to native-born individuals (Fairlie, 2009). Further, 25% of founders of new high-tech companies with more than $1 million in sales are immigrants (Wadhwa, Saxenian, Rissing, & Gereffi, 2007). Immigrants are also more likely to study in one of the STEM (science, technology, engineering, and mathematics) fields and generate innovations seeking patents (Hanson, 2011).

Yet we know little of how immigrants in the United States perceive the patent system, a key mechanism for promoting innovation. More specifically, *are institutions valued by immigrants in the United States when starting firms, and do patents help immigrants in the United States with the performance of their firms?* In 2004, 12% of the total patents were granted to individuals of Chinese or Indian descent (Kerr, 2008). Additionally, immigrants account for 24% of patent applications (Wadhwa et al., 2007). Hunt and Gauthier-Loiselle (2010) find that this patent activity by high-skilled immigrants leads to important contributions to innovation. Further, in the 1990s, patent activity by high-skilled immigrants increased the U.S. GDP per capita by around 2% (Hunt & Gauthier-Loiselle, 2010).

This study examines whether immigrants in the United States value the American patent system when starting firms, and if so, whether the patent system helps immigrants with the performance of their new firms. This issue is important not only in terms of understating decision making with regards to entrepreneurship of a large, gradually increasing subpopulation in the United States (immigrants), but also because it sheds light on government policy about innovation. Fostering innovation is one of a few public policies that could be used in order to encourage entrepreneurship in a capitalist society (Minniti, 2008). Furthermore, scholars argue that innovation has been the driving force in the emergence of the United States as the world's leading economy (Michael & Pearce, 2009). Innovation has brought competition, and competition has brought progress. An innovative spirit has never been as essential as in the current volatile business environment (Fink & Harms, 2012). In such an environment, strategic ploys used by companies are usually short-lived, and for a company to gain a sustainable competitive advantage, it needs to be innovative (Freeman, 1987, 1995; Teece, 2009).

In past decades, firms have mostly endured competition from within their own countries. However, the information age has brought globalization and a continual push towards a global economy. This global economy has resulted in a plethora of foreign nationals pursuing their educational goals through the strong American higher-education system. These immigrants have spent more time on science and engineering education compared to individuals born and raised in the United States (Borjas, 2005, 2006; Espenshade, 2001), which has resulted in a higher ratio of immigrants in science and engineering occupations in recent years (Peri & Sparber, 2011; Sana, 2010). Science and engineering represent a core driver in fostering innovative ideas in the market place, with high-skilled immigrants playing a factor. These high-skilled immigrants, with an American education, have the choice to work in their native country or stay in the United States, if an American company would sponsor them. They also have another option, and that is to start their own business in the United States. If these individuals are building innovative solutions in the companies they work for or decide to start their own companies, they can help improve the competitive nature of America's economy. Policymakers are in doubt that immigrants are helping the American economy because they feel that immigrants are net takers of jobs as opposed to net creators of jobs (Borjas, 2005; Espenshade, 2001). Based on extant research, we know that non-American entrepreneurs have a higher rate of starting firms in the United States, on a percentage basis of their respective population living in the United States, than American citizens (Hsu, Roberts, & Eesley, 2007). These firms started by non-Americans tend to generate more innovation (Saxenian, 2002a, 2002b), and have higher survival rates (Coleman, Cotei, & Farhat, 2013). However, with nontechnical entrepreneurship by non-Americans, it has been found that there is a lower survival rate than with American entrepreneurs who start non-technical firms (Georgarakos & Tatsiramos, 2009). The reason is that nontechnical entrepreneurs tend to start firms due to necessity and will thus leave the entrepreneurship ranks quickly when a wage opportunity appears. Thus, it could be that immigrants, especially high-skilled immigrants (e.g., ones who have been educated in science and engineering in the American system), can start new firms that are leading to more innovation (Saxenian, 2002b). The balance of the chapter proceeds by building the theoretical framework for testable hypotheses, outlining the research methods and data, presenting the results and discussing the implications of these results.

BACKGROUND AND THEORY DEVELOPMENT

High-Skill Versus Low-Skill Immigrants

We define an immigrant as an individual who was born in a foreign country, is not a U.S. citizen, and has moved to the United States to study or work. Immigration in the United States is a complex policy issue that has brought a continuous discussion from a variety of sources. On the one hand, the immigration discussion focuses on low-skill immigration. In other words, these are immigrants brought in to fill generally undesirable jobs that are dirty, difficult, and dangerous. Proponents of low-skill immigration argue that there is no evidence that low-skill immigrants hurt the economy and if anything they actually help it (Card, 2005; Ottaviano & Peri, 2006). Opponents argue that low-skills immigration serves as basis for the redistribution of wealth in the economy from low-skill to high-skill natives (Borjas, 1999). On the other hand, the discussion is also plentiful on whether high-skill immigration is beneficial for the country. High-skill immigrants are ones who fill more demanding jobs such as the ones related to STEM (science, technology, engineering, and mathematics) fields. Advocates of high-skill immigration argue that it stimulates innovation among native-born individuals (Hunt & Gauthier-Loiselle, 2010; Kerr & Lincoln, 2010). Opponents counter that high-skill immigrants eliminate jobs and displace native-born individuals depressing wages and hurting unemployment (Borjas, 2005; Levin, Black, Winkler, & Stephan, 2004). This study does not aim to resolve this discussion. Instead, our goal is to examine immigrants' use of the U.S. patent system and its effects on company performance as seen through the eyes of an immigrant. Our main focus, therefore, is on high-skill immigrants because, as opposed to low-skill immigrants, they start businesses and apply for patents.

High-skill immigrants enter the United States on a temporary visa, usually an F-1 visa that allows them to get their education. Once they complete their degrees, immigrants have limited options. One option is to return to their country, apply what they have learned, and in the process, advance their own or even their country's agenda. Another option is to find a job in the United States. Finding a job is difficult, however, due to the requirement that immigrants need to be sponsored by their employer in order to be issued a working visa, called H1B. This sponsorship is a costly and complicated process that in many cases turns away potential employers. What makes things even more difficult for immigrants is that there is a cap of 65,000 H1B visas issued each year. In most years, this cap is filled in a day or two. These constraints result in a number of talented and high-skilled immigrants leaving the country. Faced with the option of leaving the United States, immigrants have one last option, and that is to start a new business, which then could sponsor them to stay in the country.

Immigrant Entrepreneurs and Patenting

Extant research presents us with various views on why people patent with no consensus on what the primary driver is (Graham & Sichelman, 2008). Graham and Sichelman (2008) summarize in detail these explanations. One explanation is that patents provide greater than average returns for the patent holder. Another view is that patents are used as a tool to protect the patent holder from future patent infringement lawsuits. The next explanation is that larger organizations acquire patents as a way to provide leverage for themselves in future negotiations with competitors. Another view is that companies patent in order to increase their chances of getting acquired. Finally, patents are posited as a mechanism to reduce information asymmetry in the patentee–investor relationship. Information asymmetry occurs in this relationship because neither of the sides has perfect information on whether the patent provides useful benefits to the organization. Further, investors know little about the entrepreneur/patentee and whether the entrepreneur has their best interest at heart.

In the context of immigrant entrepreneurs and patenting, we focus on the information asymmetry view as a key determinant for how immigrants perceive patents. New firms face higher information asymmetries as opposed to established firms (Lounsbury & Glynn, 2001). When founders of new start-ups are immigrant entrepreneurs, as opposed to native entrepreneurs, information asymmetry is increased due to the lack of information about the founder. Increased information asymmetry could lead to financiers not investing in an otherwise suitable venture. It is in the best interest of immigrant entrepreneurs to reduce information asymmetry in order to increase their chances of acquiring financing. Immigrant entrepreneurs, however, have limited options in order to lower the inherent informational asymmetries due to restrictions placed by the U.S. government on noncitizens. We argue that immigrants will look to the U.S. patent system in order to send a signal to investors that they have investors' best interest at heart. Patents are validated and certified by a creditable 3rd party, the U.S. government. This certification will reduce information asymmetry and make investors more likely to invest in the immigrant entrepreneurs' venture. Because of this:

> **Hypothesis 1:** *Immigrant entrepreneurs in the United States will use patents more than American-born entrepreneurs when starting and operating their firms.*

Patents and Firm Performance

There is a plethora of research examining the relationship between patents and performance (e.g., Artz, Norman, Hatfield, & Cardinal, 2010; Hunt

& Gauthier-Loiselle, 2010; Peri & Sparber, 2011). There have been, however, mixed findings. On the one hand, we have evidence that patents do not affect firm performance of established firms (Artz et al., 2010; Deeds & Decarolis, 1999). On the other hand, patents have been found to positively impact performance of established firms (Del Monte & Papagni, 2003; Ernst, 2001). Most of the studies on the relationship between patents and performance have focused on established firms. This study examines the relationship in a new firm context and with a focus on immigrant entrepreneurs.

Immigrants are often thrust into an unfamiliar environment with very few resources at their disposal. Further, they have many restrictions placed on them by the U.S. government that further complicate the acquisition of resources. The patent system aims to provide a level of protection for firms with unique inventions essentially making some innovations inimitable (Graham & Sichelman, 2008). As inimitability of acquired resources is a key precursor to attaining a sustainable competitive advantage and having a sustainable competitive advantage is essential in ensuring high firm performance (Barney, 1991), we argue that:

Hypothesis 2: *Having a patent will be positively associated with new firm performance of immigrant entrepreneurs in the United States.*

RESEARCH METHODS

In this study, we examine whether immigrants in the United States value the American patent system, and if so, whether the patent system helps immigrants with the performance of their firms. In the previous section, we explained the reasoning why immigrants will possibly value patents more than their American counterparts (H1), and that patents will help immigrant entrepreneurs in the United States with performance of their firms (H2). To test the two hypotheses, we use data from a longitudinal study and analyze that data with multiple statistical methods.

The data set used for this study is the private version of the Kauffman Firm Survey (KFS), which is a longitudinal survey conducted by the Ewing Marion Kauffman Foundation. For the Kauffman Firm Survey (KFS), 32,469 firms were randomly sampled from Dun and Bradstreet's database of roughly 250,000 firms that started in 2004 in the United States.[2] 17,258 firms were screened for eligibility that resulted in the identification of 6,030 eligible firms (DesRoches, Mulcahy, Robb, & Shane, 2008). The KFS research team initially interviewed founders of the 4,928 firms between July 2005 and July 2006, representing a response rate of 43% from the original sampling frame after the sampling weights were applied (Ballou et al., 2007). These firms were then surveyed on an annual basis thereafter. The mode of data collection consisted of computer assisted telephone interviews (CATI), Internet, and mail.

This study uses data through 2009. Six years of data are used for the performance (growth and survival) analysis performed in this study. Six is an important number of years to examine, as most studies on failure of new firms use six years or less (e.g., Bruderl, Preisendorfer, & Ziegler, 1992; Cressy, 2006; Evans, 1987a, 1987b; Mata & Portugal, 1994; Phillips & Kirchhoff, 1989). To create the sample to be used in the analysis, we first construct the longitudinal population from the KFS data set. Out of the 4,928 firms in the baseline survey, 1,631 firms are removed from the sample due to nonresponse in follow-up years. We then remove the observations of firms that are not started as a new independent business, with zero capital in year one, no founding team members, missing data in the credit source variables, didn't use credit within six years, sold or merged, and outliers and those with obvious incorrect data entry. Firms that are purchased as an existing business, purchased as an existing franchise, or are started by means other than from scratch are excluded. This criterion entails removing 217 firms. Some firms could be started by individuals who file business tax forms to reduce tax liability rather than for business reasons. Thus, in following Astebro and Bernhardt (2003) and Bates (1990), we remove firms that have zero capital at start-up, which is measured by the total amount of assets in year one.[3] There are 381 firms with zero capital in year one. Following the same reasoning, firms that do not have founders are removed from the sample. This results in removing an additional 16 firms. In addition, firms that are sold to another firm or merged with another firm during the six-year period of the data set are removed as this represents a positive closure as opposed to a negative closure. This is the same argument made by Gimeno, Folta, Cooper, and Woo (1997) when they removed sold companies in their analysis of survival of new firms. A negative closure occurs when the firm stops operations, which represents a negative outcome and thus a failure. A positive closure of a firm is a positive outcome, and thus is not a failed firm. We are interested in equating "nonsurviving" with a negative outcome, not a positive outcome. This results in removing 122 firms. Missing data is an issue in statistical analysis, and thus firms that have any missing data in their credit source variables (dependent variables in the analysis) in any year are removed. Missing data in a longitudinal study can be due to respondents choosing not to answer a specific question in a given year or due to attrition in the respondents over time. This results in 139 firms being removed. Finally, 32 observations with incorrect data entry or outliers are removed. Incorrect data entry and outliers are removed by using the *explorer* function in SPSS and through a visual inspection of each variable used in the analyses based upon realistic levels (e.g., one firm indicates it used 50 credit sources, and another firm indicates it has revenue of

$800 million in year three when its previous and following years of revenue are minuscule). A total of 1,663 firms are removed from the longitudinal population, leaving 2,390 firms that provide valid responses in the first year (2004) of the survey and these are used in the statistical analysis.

Multiple statistical analyses are used in this study to test the two hypotheses and address the two parts of the research question. Initially, we perform *t*-tests and ANOVA analysis on the entire data set to compare the use of patents between immigrant-founded firms and American-founded firms. This analysis addresses hypothesis 1. To address hypothesis 2, we then use several additional statistical methods that effectively incorporate the annual data points in the longitudinal data on only immigrant-founded firms, including Cox proportional hazard modeling (survival analysis) and panel data regression (growth analysis). Table 7.1 is a list of the variables and how each variable is defined.

The two primary variables being examined for testing the first hypotheses are related to patents owned by the firm and the immigrant makeup of the founding team of the firm. A dichotomous variable for patents has been setup to indicate if a firm has at least one patent. The patent variable equals 1 for firms that have a patent in any of the years of operations and equals 0 otherwise. A firm is considered an immigrant firm when at least one member on the team is not an American citizen. This is a dichotomous variable, where 1 indicates the firm has at least one non-American citizen on the founding team, and 0 if all the founding team members are American citizens.

The technology level of a firm (in terms of the industry) can possibly affect whether a firm will use patents. In addition, survival and growth rates have been found to be different for firms in different technology industries. High-technology firms tend to spend a higher percentage of their funds on research and development. Research and development expenditures are difficult to finance via external means, and for new firms, equity (either internal or external) tends to be the source used (Auerswald & Branscomb, 2002; Brierley, 2001; Carpentier & Suret, 2006; Gompers & Lerner, 2001; Hall & Lerner, 2010; Lerner, 2010; Lindstrom & Olofsson, 2001). Founders in general have been found to sacrifice economic rationality for nonfinancial rewards (Cyert & March, 1963; Simon, 1955), and founders of high-technology firms tend to violate economic rationality even more so than founders of firms that are not high technology firms. Reasons include independence (Deakins & Philpott, 1994; Oakey, 1984, 1995) and location (Greenhut & Colberg, 1962). Thus, founders of high-technology firms are more willing to substitute profit maximization with nonfinancial objectives as a key goal for their firms as long as the firm does not become insolvent (Oakey & Cooper, 1989). The high-technology founder may be prepared to have his or her firm remain static or grow slowly by avoiding the use of

TABLE 7.1 List of Variables

Variable	Measurement
Patents	0 = no patents, 1 = obtained a patent within the first 6 years
Immigrant/American-founded Firm	0 = all American founders, 1 = at least one non-American citizen on the founding team
Technology Level	0 = the firm in a low technology sector, 1 = the firm is in a high technology sector
Product/Service	0 = the firm offers a service, 1 = the firm offers a product or a product/service
Variables Only Used in the Survival and Growth Analysis	
Number of Founders	The number of founders on the team
Education of Founders	Average level of education of the founders (categorical: 1–10)
Work Experience of Founders (overall)	Average number of years of overall work experience of the founders
Work Experience of Founders (start-up)	0 = if no founders have started a firm before, 1 = if at least one founder has started a firm before
Age of Founders	Average age of founders
Gender	Percentage of females on the founding team
Founder Contributed Equity	Natural log of the dollar value contributed by founders in year 1
Number of Hours Worked in Venture	Average number of hours worked by the founders each year (last year, 6 years)
Amount of Outstanding Debt	Natural log of the dollar value of debt that is outstanding for a firm each year (last year, 6 years)
Amount of New Equity	Natrual log of the dollar value of equity raised by a firm each year (last year, 6 years)
Survival	For each year (6 years): 0 = if firm is alive, 1 = if firm has closed
Growth: Revenue	Natural log of the dollar value of the revenue for a firm each year (last year, 6 years)
Growth: Employees	Natural log of the number of employees for a firm each year (last year, 6 years)

external funds (Deakins & Philpott, 1994). This attitude could affect the survival rate of high-technology firms.

For the performance variables, survival and growth, the sample includes only firms that have at least one immigrant individual on the founding team. This amounts to 187 firms. In explaining the survival and growth variables below, the figures illustrating the means and standard deviations are only for those firms, not the entire sample.

Survival

The distribution of the closure (the inverse of survival) of the firms for immigrant only firms is shown in Table 7.2, a life table indicating the number of firms that died/survived each of the six years. Of the 187 immigrant-founded firms that are in the data set, there are still 125 firms operating after year six. They are considered "censored" as the data collection period stops after year six, and they could still possibly die sometime after year six. The cumulative closure rate for firms through year six is 38.5%.

Growth

Growth is an important measurement of firm performance (e.g., Brush & Vanderwerf, 1992; Danson, 1999), but it is multidimensional (Parker, 2009). Growth can be measured by revenue, profit, number of employees, assets, and other variables. However, in this analysis, we use only revenue

TABLE 7.2 Life Table (Survival)—Only Immigrant-Founded Firms

Year	Year	Time Interval	# of firms at beginning of year	Firms that died during year	Censored at end of year	Proportion of firms at beginning of year who left during year (Hazard Function)	Firms still in operations at end of year (Survival Function)
2004	0	[0,1)	187	0	0	0.0%	100.0%
2005	1	[1,2)	187	22	0	11.8%	88.2%
2006	2	[2,3)	165	15	0	9.1%	90.9%
2007	3	[3,4)	150	12	0	8.0%	92.0%
2008	4	[4,5)	138	4	0	2.9%	97.1%
2009	5	[5,6)	134	9	125	6.7%	93.3%

and number of employees to represent growth, which have been regarded as very important indicators in past studies (e.g., Hmieleski & Baron, 2009). Sales growth tends to be the indication of performance that entrepreneurs use (Barkham, Gudgin, Hart, & Hanvey, 1996; Robson & Bennett, 2000), whereas policymakers are more interested in employment growth (Parker, 2009). New firms do not have a large amount of profit and assets during their early years, so their trends might not represent a clean representation of growth for a new firm. Prior studies indicate that employment is correlated with profit and assets (Davidsson, Delmar, & Wiklund, 2006). Thus, profit and assets might provide the same results as employment and would justify not using these variables. Growth is represented in multiple models for each one of these dimensions: revenue and number of employees.

The distribution of growth for revenue and number of employees is shown in Table 7.3. Average revenue grows from $0 before 2004 to $899,277 in 2009. Revenue grows steadily over the first 5 years before a decline in the 6th year. The decline in revenue in the sixth year could have been caused by the recession that began in December 2007. Standard deviation of revenue has the same trend over time, with the largest standard deviation being $3.924 million in 2008. The large standard deviation compared to the mean indicates that revenue is skewed. There is a small number of firms that experience a larger growth rate than the majority of firms. The average number of employees grows from 0 before 2004 to 4.616 in 2009. As with revenue, the number of employees increases for the first 5 years before declining slightly in 2009. However, unlike revenue, the rate of growth in employees slows earlier. The standard deviation increases over time to 12.794 in 2008. As with revenue, the employee variable shows that a few firms grow at a higher rate, but the majority of firms grow at a much lower rate in terms of number of employees.

Based upon past research on growth and survival of new firms, various control variables are used in the analysis, including technology level,

TABLE 7.3 Distribution of the Growth Variables—Only Immigrant-Founded Firms

	Revenue			Employees		
Year	Mean	Std Dev	*N*	Mean	Std Dev	*N*
2004	$108,639	$183,972	187	1.98	4.80	187
2005	$312,318	$555,241	165	3.55	6.14	165
2006	$441,479	$736,808	150	4.08	6.65	150
2007	$753,404	$1,973,870	138	4.14	7.78	138
2008	$935,796	$3,924,130	134	4.78	12.79	134
2009	$899,277	$3,756,770	125	4.62	12.12	125

product/service, number of founders, education of founders, work experience of founders, age of founders, gender, founder contributed equity, number of hours worked in the venture, amount of outstanding debt, and amount of new equity. Each one of these variables is explained below.

The level of technology is defined by the amount of innovation or human capital in an industry. Following Chapple, Markusen, Schrock, Yamamoto, and Yu (2004), the level of technology of a firm is defined in two ways: (a) technology employers, and (b) technology generators. Technology employers are identified as high technology if they are in an industry where the employment of these occupations exceeds three times the national average of 3.33%, or 9.98% (Chapple et al., 2004). Technology generators are identified as firms being in industries (as defined by NAICS codes) that exceed the U.S. average for both research and development expenditures per employee ($11,297) and for the proportion of full-time equivalent research and development scientists and engineers in the industry workforce (5.9%). If a firm is in an industry that is either a technology employer or technology generator, it is considered high technology. Firms in all other industries are considered low technology. This is a dichotomous variable, where 1 represents high technology and 0 represents low technology.

Product and service firms have different capital requirements that could affect whether patents are needed or required to compete effectively. The service sector has a smaller need for tangible assets than the product (manufacturing) sector, and thus would require less. Thus, based upon whether the firm is a product or service firm, it is important to control for the type of firm. A dichotomous variable has been created to determine if the firm provides a service (0) or a product (1).[4]

The size of the founding team could affect survival and growth, as some studies have found that larger founding teams have better performance (e.g., Eisenhardt & Schoonhoven, 1990). Thus, a variable indicating the number of founders for a firm in the first year of existence (2004) is calculated.

In addition, education and work experience (both overall and in startups) has been found to affect firm performance (e.g., Cooper, Gimeno-Gascon, & Woo, 1994; Gimeno et al., 1997; Parker, 2009). Education was measured on the highest level for each founder as of 2004, ranging from 1 (less than 9th grade) to 10 (professional school or doctorate). Each founder of a firm reported the number of years of work experience in the firm's industry. The average work experience of the founders is measured by summing the years of work experience of each member of the founding team and dividing by the number of members on a founding team. Each founder of a firm reported whether he or she had started a previous firm. If at least one founder had started a firm before, this variable is 1, and otherwise 0.

The age of founders could influence the survival and growth of the firm. New firms run by older founders have been found to survive longer as older founders can stave off bankruptcy or these older founders are more skilled at running a firm (e.g., Bates, 1990; Cosh & Hughes, 1994; Cressy, 2006). In addition, young founders are more likely to find more opportunities outside of their firm and voluntarily exit the venture. Finally, Hsu et al. (2007) show that founders who graduated from technology-based universities have a decreasing median age over time.

Despite the growing number of female-owned firms in the U.S., female-owned firms tend to be smaller in size and have slower growth rates than male-owned firms (e.g., Cliff, 1998; Cooper et al., 1994; Fischer, 1992; Fischer, Reuber, & Dyke, 1993). In addition, firms started by women have a lower survival rate than firms started by men (Bosma, Van Praag, Thurik, & De Wit, 2004; Robb, 2002). Some studies have shown that women tend to request less external and institutional financing (Alsos, Isaksen, & Ljunggren, 2006; Carter & Rosa, 1998) With patents a possible way to decrease the information asymmetry between entrepreneurs and external funders, and women requesting less external funds, gender could affect if a firm uses patents. To control for gender, a variable indicating the percentage of females on the founding team is computed.

The level of commitment to the venture by founders can be an important factor in determining whether a new firm survives and grows (Reynolds & Miller, 1992). For example, Reynolds (2008) says the intensity of effort is a positive indicator for venture success. Level of commitment can be measured by number of hours worked by the founders and the amount of capital initially provided to the venture at startup (Semasinge, Davidsson, & Steffens, 2011). The number of hours that founders put into the venture could affect the success of the new firm, with more hours resulting in better performance (Gatewood, Shaver, & Gartner, 1995). The variable is calculated by adding the total number of hours worked each week by all the founding team members and dividing by the number of founders.

Investment of money by a founder can possibly affect the performance of new firms. Cooper et al. (1994) find that initial financial capital affects the growth and survival of new firms, whereas Cassar (2004) finds that founders with higher household income have higher growth intentions for their firms. The amount of funds provided by the founders in the form of equity in year one is included in the model.

The amount of external funding obtained by firms can affect whether a firm survives and grows, while being a factor of being able to afford filing a patent(s). External funding can be measured by external debt and equity. Thus, we include the amount of outstanding debt and amount of new equity funding for each of the six years in the data set. These two variables are not normally distributed, and thus the natural log of these variables is used for both of them.

A summary of the means and standard deviations for the variables explained above is provided in Table 7.4. The table includes the means and standard

TABLE 7.4 Descriptive Statistics (Means and Standard Deviations)

Variable	Entire sample of firms		American-founded firms		Immigrant-founded firms	
	Mean	Std Dev	Mean	Std Dev	Mean	Std Dev
Patents	0.07	0.25	0.06	0.25	0.11	0.32
Immigrant/American-Founded Firm	0.08	0.27	0.00	0.00	1.00	0.00
Technology Level	0.43	0.50	0.43	0.49	0.51	0.50
Product/Service	0.51	0.50	0.51	0.50	0.48	0.50
Number of Founders	1.38	0.68	1.36	0.67	1.63	0.79
Education of Founders	6.38	2.02	6.31	2.00	7.12	2.08
Work Experience of Founders (overall)	12.86	10.31	13.04	10.38	10.64	9.18
Work Experience of Founders (start-up)	0.46	0.50	0.45	0.50	0.59	0.49
Age of Founders	45.29	11.12	45.34	11.11	44.60	11.21
Gender	0.29	0.41	0.29	0.42	0.25	0.35
Founder Contributed Equity	$24,411	$64,910	$22,836	$61,177	$42,943	$97,105
Founder Contributed Equity (Natural Log)	$4.42	$8.21	$4.36	$8.20	$5.13	$8.34
Number of Hours Worked	35.81	22.42	35.94	22.46	34.23	21.93
Total Outstanding Debt	$151,747	$2,280,323	$156,849	$2,372,219	$91,701	$408,497
Total Outstanding Debt (Natural Log)	–$1.75	$9.52	–$1.77	$9.51	–$1.47	$9.69
Total New Equity	$34,576	$387,212	$35,565	$402,535	$22,954	$88,230
Total New Equity (Natural Log)	–$3.14	$8.71	–$3.20	$8.67	–$2.36	$9.15

N = 2,390 firms for Entire Sample of Firms; n = 2,203 firms for American-founded firms; n = 187 firms for Immigrant-founded firms

deviations for the entire sample, firms with only American founders, and firms that have at least one immigrant individual on the founding team.

RESULTS

In this section, we report the results of the statistical analyses. As mentioned in the previous section, multiple statistical methods are used, spanning from simple statistical analyses to more sophisticated statistical analyses. We perform *t*-tests, survival analysis using Cox proportional hazard modeling, and panel data regression analyses. The statistical software *R 2.15.1* (R Development Core Team, 2012) is used for the statistical analysis. *R* is an integrated suite of software tools that can be used for data manipulation, calculation, and graphical display.[5] We use the base statistical package in *R* for *t*-tests and ANOVA analysis. The other packages we use are *survival* (Therneau, 2012) and *plm* (Croissant & Millo, 2008). The *survival* package examines and models the time it takes for events to occur. The package uses the *coxph* function to fit the Cox proportional hazard regression model. The *plm* package uses the generalized least squares estimation method to fit an econometrics-based model, panel data regression.

Common Method Bias

Before running the detail analysis, we investigate for common method bias issues since respondents provide answers for both independent/control and dependent variables. Common method variance is not an issue in the data set even though the same respondent who answered the independent variables also answered the dependent variable in the survey. When the same respondent answers the independent variable and dependent variable the association of these variables could be stronger than they really are. This could possibly bias the coefficient estimates due to common method variance (Podsakoff, MacKenzie, Lee, & Podsakoff, 2003; Podsakoff & Organ, 1986). Hence, we test for common method bias by examining the correlations and performing a Harman's single-factor test (Podsakoff & Organ, 1986). In addition, we conduct a Harman's single-factor test to test for common method variance. This is done by performing a factor analysis with principal components extraction using all of the independent and dependent variables included in the analysis. Fifteen factors have eigenvalues greater than 1, with the first factor accounting for only 11.5% of the variance. Since one factor did not explain more than 50% of the variance, this suggests that common method variance should not be a critical problem for this study.

Testing Hypothesis 1

To test hypothesis 1, which posits that firms founded by immigrants will use patents more than their American counterparts, we run a *t*-test and ANOVA analysis. Tables 7.5 and 7.6 present the results from the *t*-test and ANOVA analysis. The *t*-test results indicate a statistical difference in means of patent usage between firms founded with at least one immigrant and firms founded with only American citizens (mean difference = 0.05, $t = 2.46$, $p = 0.014$). The same is the case with the first ANOVA test, which includes only technology level and product/service variables (mean square = 0.28, $F = 4.52$, $p = 0.034$). However, when including other variables about the founding team (number of founders, education of founders, work experience of founders—overall and startup, age of founders, and gender of founders) and commitment and external financing variables, the statistical significance of the means for immigrant-founded firms becomes insignificant. This suggests that with basic analysis, hypothesis 1 is supported—that immigrant-founded firms do use patents more than their American counterparts. However, when including additional variables in the mean difference analysis, the support for hypothesis 1 is not as strong.

Survival Analysis

Survival analysis is performed using Cox proportional hazard modeling (Cox, 1972) with time-invariant variables. The time invariant analysis uses the values of the last year the firm was alive for debt, equity, and number of hours worked variables. In the survival analysis, the event that is being evaluated is closure of the firm. The dependent variable is thus coded as 0 if the firm is surviving and 1 if the firm has closed for each year. When the coefficient estimate is positive, it indicates a positive relation with closure of a firm.

Cox proportional hazard modeling is similar to multiple regression except the dependent variable is the hazard function at a given time. The baseline hazard function corresponds to the probability of firm closure when all the explanatory variables are zero. The baseline hazard function is similar to the intercept in multiple regression. The regression coefficients are proportional changes in the hazard and are estimated with the maximum likelihood method.

The results for the Cox proportional hazard modeling analysis are presented in Table 7.7. For the time invariant results, there are 183 observations with 124 firms that survive and 59 firm that close during the data collection period—there were four firms removed by the analysis due to missing data. The Wald test is statistically significant, which indicates the explanatory variables included in the model make a statistical difference in the hazard function when comparing the model with the explanatory

TABLE 7.5 *t*-Test Results

	Firms Founded with No Immigrants (n = 2,201)		Firms Founded with an Immigrant (n = 187)					
	Mean	Std. Deviation	Mean	Std. Deviation	Mean Difference	Std. Error Difference	t	Sig. (2-tailed)
Obtained at least one patent within first 6 years	0.07	0.25	0.11	0.32	0.05	0.02	2.46	0.01

N = 2,390

Note: Equal variances assumed

TABLE 7.6 ANOVA Results

Variable	Mean Square	*F*	Sig.	Mean Square	*F*	Sig.	Mean Square	*F*	Sig.
Patents	0.27	4.52	0.03**	0.06	0.97	0.33	0.06	1.00	0.32
Technology Level	2.77	45.51	<0.001***	1.61	27.08	<0.001***	1.77	29.74	<0.001***
Product/Service	4.77	78.52	<0.001***	4.37	73.54	<0.001***	4.06	68.06	<0.001***
Number of Founders				1.28	21.57	<0.001***	1.28	21.54	<0.001***
Education of Founders				1.03	17.29	<0.001***	0.99	16.64	<0.001***
Work Experience of Founders (overall)				0.03	0.44	0.51	0.00	0.06	0.81
Work Experience of Founders (start-up)				0.30	5.11	0.02*	0.31	5.26	0.02*
Age of Founders				0.01	0.20	0.65	0.01	0.14	0.71
Gender				0.64	10.81	<0.01**	0.60	10.03	<0.01**
Founder Contributed Equity							0.02	0.41	0.52
Number of Hours Worked							0.15	2.50	0.11
Total Outstanding Debt							0.02	0.34	0.56
Total New Equity							0.19	3.25	0.07+

Statistical significance: $^{+}p < 0.10$; $^{*}p < 0.05$; $^{**}p < 0.01$; $^{***}p < 0.001$
$N = 2{,}390$

TABLE 7.7 Results of Survival Analysis Using Cox Proportional Hazard Modeling

	Time Invariant			
	Coefficient	Std. Err	z	Pr(>|z|)
Patents	–1.42	0.76	–1.86	0.06⁺
Technology Level	0.00	0.28	0.02	0.99
Product/Service	0.01	0.29	0.05	0.96
Number of Founders	–0.18	0.20	–0.88	0.38
Education of Founders	–0.04	0.07	–0.59	0.56
Work Experience of Founders (overall)	–0.01	0.02	–0.45	0.65
Work Experience of Founders (start-up)	–0.10	0.28	–0.37	0.71
Age of Founders	–0.01	0.01	–0.75	0.45
Gender	0.50	0.36	1.38	0.17
Founder Contributed Equity	0.00	0.02	0.15	0.29
Number of Hours Worked	–0.01	0.01	–1.06	0.29
Total Outstanding Debt	–0.03	0.02	–1.76	0.08⁺
Total New Equity	0.06	0.01	3.83	<0.001***
R^2	0.16			
Wald Test	28.30			
df	13.00			
p-value	0.01			
N	183			
Observations deleted because of missingness	4			
Number of Closures	59			

Statistical significance: $^+p < 0.10$; $^*p < 0.05$; $^{**}p < 0.01$; $^{***}p < 0.001$

variables ($h(t)$) with the baseline model ($h_0(t)$). The explained variation statistic, R^2, is 16.2%. In the time invariant model, having a patent is negatively associated with closure and weakly statistically significant with closure (-1.41, $z = -1.85$, $p = 0.064$). Since closure (not survival) is the event that is being measured, a negative coefficient indicates that having a patent is associated with a higher probability of survival and thus decreases the probability of closure as an immigrant-founded firm ages—in other words, weakly supporting hypothesis 2. For the control variables, only total new equity is statistically significant (0.06, $z = 3.83$, $p < 0.001$).

Growth Analysis

Panel data regression is performed for a robust growth analysis on two different variables: revenue and number of employees. Panel data regression can better handle unobserved heterogeneity than OLS regression.

Table 7.8 shows the panel data regression results for both revenue and number of employees as dependent variables. Patents are not found to be statistically significant with either revenue (0.41, $t = 1.03$, $p = 0.303$) or number of employees (–0.26, $t = -0.09$, $p = 0.929$), thus suggesting that hypothesis 2 is not supported based upon growth analysis using panel data regression. However, the fixed effects model might be inconsistent per the Hausman test.

Fixed and random effects models are compared using the Hausman test to determine if we need to use a fixed effects model or a random effects model (Hausman, 1978). The null hypothesis of the Hausman test is that the errors from the equation and the estimators are not correlated. The Hausman tests indicate a statistical difference between the models—for the dependent variable being revenue ($F = 17.04$, $df = 490$, $p < 0.001$) and for the dependent variable being number of employees ($F = 4.05$, $df = 689$, $p = 0.007$). This suggests that the null hypothesis is rejected in all the Hausman tests and that the random effects model is inconsistent. Thus, the fixed effects model should be used. However, in the fixed effects model, the patent variable is not included as it is a fixed effect. Thus, it weakens the use of panel data regression for this study since the estimation of the fixed effects model can be inconsistent.

For robustness purposes, the survival and growth analyses were also performed on firms founded by Americans. The results were similar to the firms founded by immigrants, suggesting that filing a patent is not associated with the performance of new firms in general, thus supporting the findings of Artz et al. (2010) and Deeds and Decarolis (1999) in the context of new firms.

DISCUSSION AND CONCLUSION

In this study, we examined if immigrants in the United States value the American patent system, and if so, whether the patent system helps immigrants with the performance of their new firms. The results suggest that immigrants use the American patent system more than their American counterparts when building their own firm, as firms started by immigrants have a higher percentage of patents than firms founding by Americans, but that patents have mixed effects on the performance of firms founded by immigrants.

In testing hypothesis 1, the results are clear when performing the initial analysis. Firms founded with at least one immigrant individual tend to use patents more than firms that only have American individuals on the founding team, thus supporting Hunt and Gauthier-Loiselle (2010) and Peri and Sparber (2011). This was seen in the *t*-test results and the ANOVA analysis when only including the variables that pertain to patents—technology level

TABLE 7.8 Results of Growth Analysis Using Panel Data Regression (Revenue & Employees)

	Dependent Variable: Revenue				Dependent Variable: Number of Employees			
	Fixed Effects Model				Fixed Effects Model			
	Estimate	Std Err	*t*-value	Pr(>|*t*|)	Estimate	Std Err	*t*-value	Pr(>|*t*|)
Patents	0.41	0.40	1.03	0.30	–0.27	3.00	–0.09	0.93
Technology Level	–0.22	0.25	–0.87	0.38	2.89	1.67	1.73	0.08^{+}
Product/Service	–0.29	0.25	–1.17	0.24	0.56	1.01	0.56	0.58
Number of Founders	0.57	0.17	3.46	<0.001***	1.02	0.62	1.65	0.10^{+}
Education of Founders	0.03	0.06	0.42	0.68	–0.04	0.25	–0.17	0.86
Work Experience of Founders (overall)	0.02	0.01	1.50	0.13	0.02	0.06	0.26	0.80
Work Experience of Founders (start-up)	0.06	0.25	0.25	0.80	0.89	1.00	0.89	0.37
Age of Founders	–0.01	0.01	–1.13	0.26	0.01	0.05	0.13	0.90
Gender	0.22	0.36	0.60	0.55	–1.21	1.40	–0.86	0.39
Founder Contributed Equity	0.04	0.01	2.59	0.01**	0.07	0.06	1.14	0.25
Number of Hours Worked	0.02	0.00	5.88	<0.001***	0.04	0.01	2.70	0.01**
Total Outstanding Debt	0.03	0.01	4.54	<0.001***	0.08	0.03	3.07	<0.01**
Total New Equity	–0.03	0.00	–6.72	<0.001***	–0.07	0.02	–2.79	0.01**
R^2	0.33				0.04			
Adjusted R^2	0.33				0.04			
F-statistic	23.36				2.94			
df	641				863			
p-value	<0.001				<0.001			
n	162				185			
Time periods	1–6				1–6			
N	655				877			

Statistical significance: $^{+}p < 0.10$; $^{*}p < 0.05$; $^{**}p < 0.01$; $^{***}p < 0.001$
Fixed Effects Models
Unbalanced panel

and product/service. However, when including the variables that pertain to firm performance (founding team characteristics, commitment to the firm, and external funding), the mean difference between immigrant-founded firms and American-founded firms is no longer statistically significant. The statistical insignificance in patent usage in the ANOVA analysis with all the variables could be that those variables have a logical reason affecting the patent usage by immigrant-founded firms (e.g., females tend to not focus on STEM fields, and STEM fields have a higher probability of patent use), or they could only be providing noise which diminishes the statistical significance seen in the initial analysis.

We argued in the theory development section of this chapter that the reason we would see a mean difference in patent usage is because of the information asymmetry between entrepreneurs/firms and external stakeholders. To further try to confirm this logical reasoning, we performed additional analysis to find out if immigrant firms use external funding more—since obtaining external funding requires reducing the information asymmetry between the external funders and the entrepreneur/firm. Immigrant-founded firms did not have a statistically significant difference in the amount of new equity or outstanding debt in any of the six years of the data set.[6] Although this is not the only test that would need to be done to confirm the information asymmetry argument, it does weaken our argument that the reason that immigrants use the American patent system is to diminish the information asymmetry, at least when obtaining external funding.

For the patent-performance relationship we used a resource-based view argument, saying that patents are a resource that is valuable, rare, and inimitable, and that can be organized around to provide a sustainable competitive advantage that would be evident in firm performance. This was posited in hypothesis 2, which pertained to the relationship between patent usage and firm performance of immigrant-founded firms. Firm performance was measured by survival, revenue growth, and employee growth. In the survival analysis, there was weak statistical support that immigrant-founded firms that have a patent tend to survive longer than immigrant-founded firms that don't. Thus, patents could possibly be a resource that provides a temporal competitive advantage that is evident in survival. However, patent usage by immigrant-founded firms didn't see a statistically significant association with revenue and employee growth. This is also the case with firms founded by Americans.

Implications for Policy

In addressing immigration in the United States, it becomes important for policymakers to see if the patent system is a mechanism to entice high-skilled immigrants to stay in the United States and build their company, especially since fostering innovation is one of a few public policies that could be used in order to encourage entrepreneurship in a capitalist society (Minniti, 2008). Employee growth might be something that policymakers would be interested in when deciding on the immigration law for entrepreneurs in the United States. The statistical significance was weak for employee growth, but that could be due to the small sample size of immigrant firms (only 187 firms). If there was a larger sample size, the statistical significance could be stronger and would suggest that immigrants who use patents hire more employees than immigrants who don't use patents. Thus, policymakers could use patent criteria for immigrants wanting to stay in the United States, as they create more jobs than immigrants who do not use the American patent system.

However, policymakers will need to be sure not to rely too heavily on the patent system for all firms. If used wrongly, the patent system can possibly deter innovation—for example, with software development. Patent trolls could possibly take advantage of the system and stop new firms from developing new innovations that can benefit society. If the patent system stifles innovation, then it would also eliminate one competitive advantage that new firms have access to. In addition, immigrants could be using the patent system to fall back upon to deal with uncertainty in new firms.

Directions for Further Research

The findings in this study suggest that immigrants who start companies in the United States value patents. However, we do not find evidence that the general use of patents affects the performance of new firms. It seems patents are used for different purposes in a new firm. It may be that some patents are the foundation of the firm's product(s), whereas other patents may be ancillary to the firm's core product(s). In addition, the patent may be used for protectionism (to protect from being sued by competitors). Thus, further research needs to examine the decisions of immigrant entrepreneurs and determine how patents are being used. Such analysis could help delineate the relationship between the use of patents and new firm growth and survival rates.

Limitations of the Study

There are several limitations of the study. The sample size of the immigrant-founded firms for the performance analysis was small (only 187 firms). A larger sample size of immigrant-founded firms could possibly improve the statistical significance of the performance analysis, especially when examining the employee growth. Many firms that have an immigrant on the founding team might not have been included in the data set as they might not have wanted to be known by the government—they wanted to keep a low profile as they could be in the United States illegally. This could be especially the case for firms that have a lead founder as an immigrant. Thus, the results in this study could be missing an entire type of immigrant-founded firms.

NOTES

1. The authors are grateful to the Ewing Marion Kauffman Foundation for financial support and access to the Kauffman Firm Survey data.
2. The United States Small Business Association (SBA) recorded 628,917 firm births in the USA in 2004 (SBA, 2007, p. 309). Hence, the Dun and Bradstreet database does not include all of the firms that are started in a given year.
3. Assets include cash, accounts receivables, equipment, land/buildings, vehicles, other owner property, and other miscellaneous assets.
4. If a firm provides both a product and a service, it is considered a product firm. This is done based upon the reasoning of capital requirements for product firms. If a firm is a product firm and decides to add a service, the capital requirements don't change drastically, unlike a service firm that decides to add a product.
5. Source: http://cran.r-project.org/doc/manuals/R-intro.html
6. Results of the external funding analysis are not included in the study due to space constraints but can be provided upon request.

REFERENCES

Alsos, G. A., Isaksen, E. J., & Ljunggren, E. (2006). New venture financing and subsequent business growth in men- and women-led businesses. *Entrepreneurship Theory and Practice, 30*(5), 667–686.

Anderson, S., & Platzer, M. (2006). *American made: the impact of immigrant entrepreneurs and professionals on US competitiveness.* Arlington, VA: National Venture Capital Association study.

Artz, K. W., Norman, P. M., Hatfield, D. E., & Cardinal, L. B. (2010). A longitudinal study of the impact of R&D, patents, and product innovation on firm performance. *Journal of Product Innovation Management, 27*(5), 725–740.

Astebro, T., & Bernhardt, I. (2003). Start-up financing, owner characteristics, and survival. *Journal of Economics and Business, 55*(4), 303–319.

Auerswald, P. E., & Branscomb, L. M. (2002). Start-ups and spin-offs: Collective entrepreneurship between invention and innovation. *The Journal of Technology Transfer, 28*(3), 227–239.

Ballou, J., Barton, T., DesRoches, D., Potter, F., Zhao, Z., Santos, B., & Sebastian, J. (2007). *Kauffman Firm Survey (KFS) baseline methodology report.* Kansas City, MO: Kauffman Foundation.

Barkham, R., Gudgin, G., Hart, M., & Hanvey, E. (1996). *The determinants of small firm growth: An inter-regional study in the United Kingdom 1986–1990.* London, England: Jessica Kingsley Publishers.

Barney, J. B. (1991). Firm resources and sustained competitive advantage. *Journal of Management, 17*(1), 99–120.

Bates, T. (1990). Entrepreneur human capital inputs and small business longevity. *The Review of Economics and Statistics, 72*(4), 551–559.

Borjas, G. J. (1999). *Heaven's Door: Immigration Policy and the American Economy.* Princeton: Princeton University Press.

Borjas, G. J. (2005). The labor-market impact on high-skill immigration. *The American Economic Review, 95(2),* 56–60.

Borjas, G. J. (2006). *Immigration in high-skill labor markets: The impact of foreign students on the earnings of doctorates.* NBER Working Paper. National Bureau of Economic Research.

Bosma, N., Van Praag, C. M., Thurik, R., & De Wit, G. (2004). The value of human and social capital investments for the business performance of startups. *Small Business Economics, 23*(3), 227–236.

Brierley, P. (2001). The financing of technology-based small firms: A review of the literature. *Bank of England Quarterly Bulletin, 41*(1), 64–83.

Bruderl, J., Preisendorfer, P., & Ziegler, R. (1992). Survival chances of newly founded business organizations. *American Sociological Review, 57*(2), 227–242.

Brush, C. G., & Vanderwerf, P. A. (1992). A comparison of methods and sources for obtaining estimates of new venture performance. *Journal of Business Venturing, 7*(2), 157–170.

Card, D. (2005). *Is the new immigration really so bad?* NBER Working Paper No. 11547.

Card, D. (2009). Immigration: How immigration affects U.S. cities. In R. P. Inman (Ed.), *Making cities work* (pp. 158–200). Princeton, NJ: Princeton University Press.

Carpentier, C., & Suret, J.-M. (2006). Some evidence of the external financing costs of new technology-based firms in Canada. *Venture Capital, 8*(3), 227–252.

Carter, S., & Rosa, P. (1998). The financing of male- and female-owned businesses. *Entrepreneurship & Regional Development, 10*(3), 225–241.

Cassar, G. (2004). The financing of business start-ups. *Journal of Business Venturing, 19*(2), 261–283.

Chapple, K., Markusen, A., Schrock, G., Yamamoto, D., & Yu, P. (2004). Gauging metropolitan "high-tech" and "i-tech" activity. *Economic Development Quarterly, 18*(1), 10 –29.

Cliff, J. E. (1998). Does one size fit all? Exploring the relationship between attitudes towards growth, gender, and business size. *Journal of Business Venturing, 13*(6), 523–542.

Coleman, S., Cotei, C., & Farhat, J. (2013). A resource-based view of new firm survival: New perspectives on industry and exit route. *Journal of Developmental Entrepreneurship, 18,* 1–25.

Cooper, A. C., Gimeno-Gascon, J., & Woo, C. Y. (1994). Initial human and financial capital as predictors of new venture performance. *Journal of Business Venturing, 9*(5), 371–395.

Cosh, A., & Hughes, A. (1994). Size, financial structure and profitability: UK companies in the 1980s. In A. Hughes & D. J. Storey (Eds.), *Finance and small firms* (pp. 18–63). London, England: Routledge.

Cox, D. R. (1972). Regression models and life-tables. *Journal of the Royal Statistical Society. Series B (Methodological), 34*(2), 187–220.

Cressy, R. C. (2006). Why do most firms die young? *Small Business Economics, 26*(2), 103–116.

Croissant, Y., & Millo, G. (2008). Panel data econometrics in R: The plm package. *Journal of Statistical Software, 27*(2), 1–43.

Cyert, R. M., & March, J. G. (1963). *A behavioral theory of the firm.* Englewood Cliffs, NJ: Prentice Hall.

Danson, M. W. (1999). Debates and reviews. *Regional Studies, 33*(1), 63–72.

Davidsson, P., Delmar, F., & Wiklund, J. (Eds.). (2006). *Entrepreneurship and the growth of firms.* Camberley, Surrey, England: Edward Elgar.

Deakins, D., & Philpott, T. (1994). Comparative European practices in the finance of new technology entrepreneurs in the United Kingdom, Germany and Holland. In R. P. Oakey (Ed.), *New technology-based firms in the 1990s* (pp. 93–103). London, England: Paul Chapman Publishing.

Deeds, D. L., & Decarolis, D. M. (1999). The impact of stocks and flows of organizational knowledge on firm performance: An empirical investigation of the biotechnology industry. *Strategic Management Journal, 20*(10), 953–968.

Del Monte, A., & Papagni, E. (2003). R&D and the growth of firms: Empirical analysis of a panel of Italian firms. *Research Policy, 32*(6), 1003–1014.

DesRoches, D., Mulcahy, T. M., Robb, A. M., & Shane, S. A. (2008). Kauffman Firm Survey: Data overview. Retrieved from http://papers.ssrn.com/sol3/papers.cfm?abstract_id=1024713

Eisenhardt, K. M., & Schoonhoven, C. B. (1990). Organizational growth: Linking founding team, strategy, environment, and growth among US semiconductor ventures, 1978–1988. *Administrative Science Quarterly, 35*(3), 504–529.

Ernst, H. (2001). Patent applications and subsequent changes of performance: Evidence from time-series cross-section analyses on the firm level. *Research Policy, 30*(1), 143–157.

Espenshade, T. J. (2001). High-end immigrants and the shortage of skilled labor. *Population Research and Policy Review, 20*(1), 135–141.

Evans, D. S. (1987a). Tests of alternative theories of firm growth. *The Journal of Political Economy, 95*(4), 657–674.

Evans, D. S. (1987b). The relationship between firm growth, size, and age: Estimates for 100 manufacturing industries. *The Journal of Industrial Economics, 35*(4), 567–581.

Fairlie, R. (2009). *Kauffman Index of Entrepreneurial Activity: 1996–2008.* Kansas City, MO: Ewing Marion Kauffman Foundation.

Fink, M., & Harms, R. (2012). Contextualizing the relationship between self-commitment and performance: environmental and behavioural uncertainty in (cross-border) alliances of SMEs. *Entrepreneurship & Regional Development, 24*(3–4), 161–179.

Fischer, E. M. (1992). Sex differences and small-business performance among canadian retailers and service providers. *Journal of Small Business and Entrepreneurship, 9*(4), 2–13.

Fischer, E. M., Reuber, A. R., & Dyke, L. S. (1993). A theoretical overview and extension of research on sex, gender, and entrepreneurship. *Journal of Business Venturing, 8*(2), 151–168.

Freeman, C. (1987). *Technology policy and economic performance: Lessons from Japan.* London, England: Pinter Publishers.

Freeman, C. (1995). The "National System of Innovation" in historical perspective. *Cambridge Journal of Economics, 19*(1), 5–24.

Gatewood, E. J., Shaver, K. G., & Gartner, W. B. (1995). A longitudinal study of cognitive factors influencing start-up behaviors and success at venture creation. *Journal of Business Venturing, 10*(5), 371–391.

Georgarakos, D., & Tatsiramos, K. (2009). Entrepreneurship and survival dynamics of immigrants to the US and their descendants. *Labous Economics, 16*(2), 161–170.

Gimeno, J., Folta, T. B., Cooper, A. C., & Woo, C. Y. (1997). Survival of the fittest? Entrepreneurial human capital and the persistence of underperforming firms. *Administrative Science Quarterly, 42*(4), 750–783.

Gompers, P. A., & Lerner, J. (2001). The venture capital revolution. *The Journal of Economic Perspectives, 15*(2), 145–168.

Graham, S. J., & Sichelman, T. (2008). Why do start-ups patent? *Berkeley Technology Law Journal, 23*(3), 1064–1092.

Greenhut, M. L., & Colberg, M. R. (1962). *Factors in the location of Florida industry.* Tallahassee, FL: Florida State University.

Hall, B. H., & Lerner, J. (2009). The financing of R&D and innovation. In B. H. Hall & N. Rosenberg (Eds.), *Handbook of the economics of innovation* (vol. 1, pp. 609–639). Amsterdam, NL: Elsevier/North Holland.

Hanson, G. H. (2011). *Immigration, productivity, and competitiveness in American industry.* Washington, DC: AEI.

Hausman, J. A. (1978). Specification tests in econometrics. *Econometrica, 46*(6), 1251–1271.

Hmieleski, K. M., & Baron, R. A. (2009). Entrepreneurs' optimism and new venture performance: A social cognitive perspective. *Academy of Management Journal, 52*(3), 473–488.

Hsu, D. H., Roberts, E. B., & Eesley, C. E. (2007). Entrepreneurs from technology-based universities: Evidence from MIT. *Research Policy, 36,* 768–788.

Hunt, J., & Gauthier-Loiselle, M. (2010). How much does immigration boost innovation? *American Economic Journal: Macroeconomics, 2*(2), 31–56.

Kerr, W. R. (2008). The ethnic composition of US inventors. *HBS Working Paper 08-006.*

Kerr, W. R., & Lincoln, W. F. (2010). *The supply side of innovation: H1-B visa reforms and U.S. ethnic inventions.* NBER Working Paper No. 15768.

Lerner, J. (2010). *Innovation, entrepreneurship and financial market cycles.* OECD Science, Technology and Industry Working Papers.

Levin, S. G., Black, G. C., Winkler, A. E., & Stephan, P. E. (2004). Differential employment patterns for citizens and non-citizens in science and engineering in the United States: Minting and competitive effects. *Growth and Change, 35*(4), 456–475.

Lindstrom, G., & Olofsson, C. (2001). Early stage financing of NTBFs: An analysis of contributions from support actors. *Venture Capital, 3*(2), 151–168.

Lounsbury, M., & Glynn, M. A. (2001). Cultural entrepreneurship: Stories, legitimacy, and the acquisition of resources. *Strategic Management Journal, 22*, 545–564.

Mata, J., & Portugal, P. (1994). Life duration of new firms. *Journal of Industrial Economics, 42*(3), 227.

Michael, S. C., & Pearce II, J. A. (2009). The need for innovation as a rationale for government involvement in entrepreneurship. *Entrepreneurship & Regional Development, 21*(3), 285–302.

Minniti, M. (2008). The role of government policy on entrepreneurial activity: Productive, unproductive, or destructive? *Entrepreneurship Theory and Practice, 32*(5), 779–790.

Oakey, R. P. (1984). Finance and innovation in British small independent firms. *Omega, 12*(2), 113–124. doi:10.1016/0305-0483(84)90030-6

Oakey, R. P. (1995). *High-technology new firms: Variable barriers to growth.* London, England: Paul Chapman.

Oakey, R. P., & Cooper, S. Y. (1989). High technology industry, agglomeration and the potential for peripherally sited small firms. *Regional Studies, 23*(4), 347–360.

Ottaviano, G., & Peri, G. (2006). *Rethinking the effects of immigration on wages.* NBER Working Paper No. 12497.

Parker, S. C. (2009). *The economics of entrepreneurship.* Cambridge, England: Cambridge University Press.

Peri, G., & Sparber, C. (2011). Highly educated immigrants and native occupational choice. *Industrial Relations: A Journal of Economy and Society, 50*(3), 385–411.

Phillips, B. D., & Kirchhoff, B. A. (1989). Formation, growth and survival; Small firm dynamics in the U.S. economy. *Small Business Economics, 1*(1), 65–74.

Podsakoff, P. M., MacKenzie, S. B., Lee, J.-Y., & Podsakoff, N. P. (2003). Common method biases in behavioral research: A critical review of the literature and recommended remedies. *Journal of Applied Psychology, 88*(5), 879.

Podsakoff, P. M., & Organ, D. W. (1986). Self-reports in organizational research: Problems and prospects. *Journal of Management, 12*(4), 531.

R Development Core Team. (2012). *R: A language and environment for statistical computing.* Vienna, Austria: R Foundation for Statistical Computing. Retrieved from http://www.R-project.org

Reynolds, P. D. (2008). *Entrepreneurship in the United States: The future is now.* New York, NY: Springer.

Reynolds, P. D., & Miller, B. (1992). New firm gestation: Conception, birth, and implications for research. *Journal of Business Venturing, 7*(5), 405–417.

Robb, A. M. (2002). Entrepreneurial performance by women and minorities: The case of new firms. *Journal of Developmental Entrepreneurship, 7*(4), 383.

Robson, P. J., & Bennett, R. J. (2000). SME growth: The relationship with business advice and external collaboration. *Small Business Economics, 15*(3), 193–208.

Sana, M. (2010). Immigrants and natives in US science and engineering occupations. *Demography, 47*(3), 801–820.

Saxenian, A. L. (2002a). *Local and global networks of immigrant professionals in Silicon Valley.* San Francisco, CA: Public Policy Institute of California.

Saxenian, A. L. (2002b). Silicon Valley's new immigrant high-growth entrepreneurs. *Economic Development Quarterly, 16*(1), 20–31.

Semasinge, D., Davidsson, P., & Steffens, P. R. (2011). Nascent venture performance: Linking novelty of venture ideas and commitment of firm founders as predictors. In A. Maritz (Ed.), *Proceedings of the 8th AGSE International Entrepreneurship Research Exchange* (pp. 597–607). Melbourne, Victoria, Australia.

Simon, H. A. (1955). A behavior model of rational choice. *Quarterly Journal of Economics, 69*(1), 99–118.

Small Business Administration (SBA). (2007). *The small business economy for data year 2007.* Washington, DC: Author. Retrieved from http://archive.sba.gov/advo/research/sb_econ2009.pdf

Teece, D. J. (2009). *Dynamic capabilities and strategic management: Organizing for innovation and growth.* Oxford, MA: Oxford University Press.

Therneau, T. (2012). *survival—A Package for Survival Analysis in S* (Version R package version 2.36-14) [R].

Wadhwa, V., Saxenian, A., Rissing, B., & Gereffi, G. (2007). *America's new immigrant entrepreneurs.* Kansas City, MO: Kauffman Foundation Report.

CHAPTER 8

UNIVERSITY RESEARCH PARKS AND THEIR IMPACT ON TECHNOLOGY TRANSFER ACTIVITIES

Jasmina Berbegal-Mirabent
Universitat Internacional de Catalunya, Barcelona

Inés Alegre
IESE Business School—Universidad de Navarra

Adrián Guerrero
Universitat Internacional de Catalunya, Barcelona

ABSTRACT

University research parks (URPs) are technology-based geographical enclaves placed in the surroundings of a university campus that benefit from the knowledge generated in the university and its research centers. In URPs, information flows from academia to industry, accelerated by a physical proximity.

Universities started building up these spaces as a result of different policies designed to foster knowledge transfer activities and spur university–industry collaborations. In Spain, this movement took place in the 1990s, and quite rapidly universities all over the Spanish territory got interested in this policy instrument as a way to foster innovation and attract R&D activity.

Technological Innovation Networks, pages 167–192

Despite literature on science parks being quite abundant, there is a theoretical gap in the mechanisms that drove the creation of these infrastructures. This chapter aims at shedding some light on this matter. First, based on the literature reviewed and the evidence of the practices followed by several URPs, we propose that URPs can be created following two different strategies, depending whether the concentration of innovative activities is based on the presence or the absence of a deliberated creation scheme. Following the preliminary work of Berbegal-Mirabent, Martín, and Solé (2010) we expand on the differences between *planned* and *unplanned parks.*

Second, an empirical study is conducted in order to determine the impact of the creation strategy of the park on technology transfer activities—namely patents, spin-offs, and R&D contracts. Taking the Spanish public higher education system and using data for the period 2004–2011, we examine the determinants of university technology transfer outputs. Then, using cluster analysis, universities are grouped. A detailed description of each group is provided. Results indicate that planned parks are natural drivers for the emergence of new ventures and facilitate the establishment of R&D contracts. On the other hand, our findings reveal that having a science park (either planned or unplanned) positively contributes to increase the amount of patents registered.

University research parks (URPs) are a particular type of science park (SP) that respond to the task of adding spaces and clusterization options to the research and valorization strategy of universities. In URPs, information flows from academia to business, accelerated by geographical proximity. Their distinctive features are mainly the logic of the location of their research centers, spin-offs, and the specificity of the promotion units and services.

URPs constitute a potentially important mechanism for university technology transfer (Leyden, Link, & Siegel, 2008). Today there are over 170 URPs in North America, the UK Science Park Association (UKSPA) reports more than 50 URPs, and more than 100 can be found in Asia. All these figures indicate that URPs represent an international phenomenon.

The growth in the number of URPs has spurred an important debate on how these strategic enclaves of innovation and cutting-edge research stimulate growth and regional economic development (Link & Scott, 2003). Particularly, literature has addressed the study of SPs from different perspectives. There is a first stream of research which includes those studies that use economic, geographical, management or public policy approaches to analyse the performance growth of URPs (Bigliardi, Dormio, Nosella, & Petroni, 2006; Link & Scott, 2006). A second bulk of studies comprises those that focus on the firm-level choice decision to locate on a URP (Leyden et al., 2008) and those comparing the economic performance of firms located inside the park with out-of-park firms (Löfsten & Lindelöf, 2001; Siegel, Westhead, & Wright, 2003; Squicciarini, 2008). Finally, a third group of studies analyse the impact of URPs in regional economic development (Chen & Kenney, 2007; Hansson, 2007).

However, despite the huge interest among policymakers and industry leaders in identifying best practices in the strategy formulation and operation of such parks, the literature is inconclusive. Particularly, two questions are still to be answered: Is the concentration of innovative activities based on the presence or the absence of a deliberated creation scheme? What are the implications of each strategy in terms of university technology transfer outputs?

As for the first question, following the preliminary work of Berbegal-Mirabent et al. (2010), we propose that there are two main strategies, distinguishing between *planned parks* and *unplanned parks*. On the one hand, *planned parks* are the result of the desire to create a URP from scratch in an attempt to provide the university with a more dynamic environment. The university is provided with one or more buildings with the appropriate spaces, infrastructures, and services to attract research groups, promoting their establishment or growth but also inducing firms and knowledge-based companies to join the project. On the other hand, *unplanned parks* follow the logic of institutionalizing the endogenous character of a dynamic environment that emerged spontaneously in the surroundings of the university. This approach first begins by offering to existing research groups at the university basic service coverage for their daily activities. As these demands get sophisticated, external firms begin to show interest in being placed there also. Consequently, specialized management units are required, leading to the formal constitution of the park.

Regarding the second question, it is widely acknowledged that URPs facilitate knowledge exchange between academics and businesses, increasing the probabilities to conduct applied research and commercialize research results. However, to the authors' knowledge, there is no empirical evidence on the real effect that URPs have on the quantity and quality of technology transfer outputs of their parent organization (the university). In this respect, we hypothesize that the effect is going to be positive, although our intuition suggests that universities may follow different patterns depending on the creation strategy of the park. An empirical study based on the analysis of the Spanish public higher education system is conducted to verify our hypothesis.

Aiming at shedding some light on these unresolved issues, the remaining of the chapter is organized as follows. First, we summarise the historical retrospective of SPs. We then focus on URPs, their distinctive features and the creation strategies. Next, we explain the methodological approach used to test the influence of URPs on university technology transfer outputs. Next, results are presented and discussed. The chapter ends with the conclusions and implications for future research.

HISTORICAL RETROSPECTIVE OF SCIENCE PARKS

The origin of SPs can be found in the United States in 1950 aiming at creating large industrial complexes of high-tech enterprises under the supervision of universities and the industry.

The first example can be found in the Silicon Valley. The initiative was led by Frederick Terman and William Shockley, professors from Stanford University, who warned that a large area of the university was underexploited, becoming ideal for real estate development. They then started promoting a program that encourages students to undertake research and doctoral theses with commercial applications, and provided them with venture capital. To date, approximately 60% of the revenues of companies from the Silicon Valley come from firms created by students and teachers.

In 1951, the project of the Silicon Valley was boosted thanks to the creation of the Stanford Industrial Park, a series of small industrial units that were made available for companies to rent at low cost for the development of new technologies. In 1954, the Honors Cooperative Program was created, allowing the employees of these companies to obtain a university degree by undertaking part-time studies. These companies signed five-year agreements, paying a double tuition fee per student to cover expenses. In the mid-1950s, the "valley" was in a nascent but growing stage, and in the early 1970s, the number of corporations engaged in the production of semiconductors increased considerably, companies that, in turn, supplied their counterpart firms that were working in the computer field area.

Countries, regions, municipalities, universities, and managers have tried to copy and replicate the singularities of the Silicon Valley. The design and implementation of similar initiatives in different forms and processes have led to the creation of SPs all over the globe.

In Europe, SPs emerged in the 1960s, being mainly established in areas near college campuses. Under the tutelage of university departments or R&D university laboratories, European SPs started promoting industrial development and innovation by creating new products and improving production processes and engineering techniques.

The SP promoted by the University of Cambridge (U.K.) and the Sophia Antipolis Science Park (France) were two of the pioneers. It was in such environments that ties between academy and industry started to flourish, resulting in collaborative partnerships that materialize in the form of R&D contracts, licensing agreements, or consulting services among others. Since the late 1970s and during the 1980s, this collaborative model expanded.

Over the years, both local and regional governments as well as an increasing number of companies demanded a stronger presence in SPs. These demands require SPs to expand their current infrastructures and offer a wider portfolio of services for firms located either inside or outside the park. This growth, however, requires of larger spaces. As a result, SPs began being located not only in universities but also in unexploited terrains and abandoned lands (Löfsten & Lindelöf, 2001; MacDonald, 1987).

Since the 1990s, SPs can almost be found everywhere and in a wide variety of environments and forms, but in all cases they are built in pursuit of promoting innovation through triple helix collaborations (Chorda, 1996; Leydesdorff & Etzkowitz, 1996).

Not all parks follow the same pattern. Depending on the location and the scope, we can distinguish among science parks, research parks, technology parks, business parks, technological districts/poles, and technopolises. Although differences exist, they are all popularly referred to as "science parks." They have several common features:

1. They are related and associated with an expert technological center (usually a university).
2. They promote the creation and development of technology-based companies through incubation and spin-off processes.
3. They provide advanced support services to those institutions located inside.
4. They facilitate technology transfer processes among the different stakeholders, accelerating the commercialization of the results.
5. They are committed with the development of their regions by stimulating competition, quality, and innovation among their associated institutions and businesses.
6. They require the management of spaces and resources by a specialized professional team.

UNIVERSITY RESEARCH PARKS

Defining a URP

The entrepreneurial university assumes that the market is a driving force that encourages external collaborations with industry, strengthening the academic performance and attracting additional non-public funding and resources (Clark, 1998). According to Etzkowitz, Webster, Gebhardt, and Cantisano Terra (2000), the entrepreneurial university transforms ideas into tangible activities, capitalizes knowledge, and creates new firms and services. It is able to deal with risk transforming knowledge into economic activity while favoring regional development.

One of the core elements that characterize an entrepreneurial university is its capacity to narrow the gap between academic research and businesses. The natural consequence of bridging these two worlds together is the creation of specific research infrastructures such as university research parks that exponentially multiply the opportunities for the development of R&D activities (Löfsten & Lindelöf, 2002).

In a URP, the university acquires a prominent role, acting as the main driver for knowledge creation and knowledge transfer (Markman, Phan,

Balkin, & Gianiodis, 2005; Phan, Siegel, & Wright, 2005). Leyden et al. (2008) state that the geographical concentration of a technological offer (universities and research groups), a technological demand (private and public entities requiring university's knowledge and expertise) and the support of the public administration, is what encourages the establishment and formalization of URPs.

According to the Association of University Research Parks (AURP), a *research park* is an expression mainly used in the United States, whereas in Canada, Europe, Asia, and Latin America other expressions are preferred such as science park or technology park. For the purpose of this study, we define a URP as a *property-based venture that has existing or planned land and buildings designed primarily for private and public research and development facilities, high technology and science based companies, and support services.* Its goal consists in *promoting research and development by the university in partnership with industry, assisting in the growth of new ventures, and promoting economic development* while *aiding the transfer of technology and business skills between the university and industry tenants.* This definition is in accordance with that of Link and Scott (2003, p. 44), who define URPs as "clusters of technology-based organizations placed in the surroundings of a university campus in order to benefit from the knowledge generated in the university and its research centres."

Creation Strategies

Previous Studies

Under the perspective of a territorial concentration of agents involved in science-based activities, there are some studies aiming at explain how URPs have been established. Nevertheless, the existing literature presents a theoretical gap concerning the strategies in the creation of such spaces (Link & Scott, 2007).

Employing patterns of growth and the territorial organization of activities, Cooke and Morgan (1994) and Braczyk and Cooke (1998) distinguished three models that help explain the origins of the concentration of innovative activity in a given territory: the "directive," the "spontaneous," and the "networking." The former is evidenced by urban or metropolitan planned actions where efforts are directed to concentrate high-tech activities through infrastructure planning, usually guided by the intervention of industrial policy. Examples of this type include Sophia Antipolis or the Grenoble ZIRST. The spontaneous model is the result of an unplanned spatial concentration of innovative activity originated by the sum of independent initiatives coming from actors located in a particular area under the umbrella of local and R&D centres. Silicon Valley and the Oxford Science Park clearly exemplify this model. Lastly, the third model provides the territory with a set of institutions conducting both basic and applied research as well as of industrial firms that operate on the basis of a hierarchical and divided innovative work (e.g., Baden-Württemberg).

In all models, the construction of the spatial vision prevails. Consequently, to some extent, the location of a URP seems not to be accidental, since URPs are settled in areas where resources are available and preconditions are favorable. Differences lie in the desire to create something from new or to exploit something existing by undertaking a formalization process.

URPs' birth is influenced by many additional factors besides those strictly related with the technological capacity and the existence of an industrial heritage. Transportation and communication infrastructures, climate, quality of the workforce, private and public funding, and so on complicate any effort in determining why, how, and where URPs are created. Moreover, according to Bigliardi et al. (2006), science parks' mission and strategy may change over time; they are usually redefined once the park comes into operation. The rationale behind this lies in the fact that only after its initial stages is it possible to evaluate the strengths and the weaknesses of the use of the spaces and the commitment of the different stakeholders.

Koh, Koh, and Tschang (2005) suggest that growth mechanisms, technological capabilities, and the global role and market integration are critical aspects that can help in the understanding of URPs. Growth mechanisms refer to those factors and capabilities that support the growth of the park over time. Technological capabilities are concerned with R&D capabilities and strengths in the creation of competitive advantage. Finally, the global role and market integration address the linkages and integration degree in regional and global markets.

From the above, it can be inferred that the concentration of innovative activities can be based on the presence or the absence of a deliberated creation scheme—that is, creating something from scratch, or exploiting something that already exists but undertaking a formalization process. The strategy chosen will undoubtedly influence the future of the park.

Planned Versus Unplanned Parks

In a planned park, universities are provided with one or more buildings with the appropriate services aimed at attracting research groups and firms to locate inside. The park itself is specifically created for generating a more dynamic environment. However, there is no tangible evidence that this goal will be achieved and that the buildings and spaces that would have been eventually created are going to be used. Thus, one of the main difficulties is the ability of attracting a critical mass of scientific talent and knowledge-based companies. A well-developed marketing strategy is indeed needed to ensure the success of the park.

It is necessary to remark that planned parks are built based on an internal design that has been drawn prior to starting with the activity. It is for that reason that unexpected obstacles may arise when implementing it. Moving from a paper proposal to a 3D reality is a very complex process that requires a powerful marketing strategy able to attract research groups, knowledge-based companies, venture capital companies, and SMEs to move into the park. Successes and failures during the project may lead to modifications in the strategy, expansions, and even changes in the relationships between the stakeholders.

Planned parks are usually products of regional governmental initiatives in an agreement with a university. In order to convince companies to locate there, the initial stages are mainly concentrated in attracting renowned R&D firms. In this respect, evidence suggests that better results are achieved if firms that are currently collaborating with the university should be the first to be invited to join the project. The scalability and growth of the park will be achieved over time.

As with any intended action on the territory, parks created following this strategy are powerful policy instruments that contribute to regional and economic development. By using housing as an urban proposal, URPs help to create a society of entrepreneurs and innovative corporations. An important aspect to consider in the decision of where to locate a URP is the current provision of general services in the area, such as the existence of modern IT services, transport facilities, proximity to airports, and so on. The quality of communication networks is paramount in order to attract potential dwellers.

A very different strategy is the one followed by unplanned parks. In this case, parks are created as a result of institutionalizing an endogenous dynamic environment, characterized by high quality services that boost research and technology transfer activities. Aiming at improving the externalities and the interrelations between the different stakeholders located in the same geographical enclave, parks are then formalized and specific units are created in order to manage all the activities that emerged spontaneously.

Contrary to planned parks, unplanned parks require the existence of a critical research mass prior to the formalization of the park. This includes highly active research groups, incubator spaces to host spin-offs, and advanced services for R&D promotion and valorization.

Unplanned parks are endogenous in their nature and better suited for universities with increasing levels of knowledge creation and knowledge transfer. As discussed in above, universities presenting these features are those that respond to the "entrepreneurial" typology defined by Clark (1998)—that is, universities that in a given moment of their evolution need to formalize all the activities through the establishment of a management body capable of organizing all the initiatives and providing appropriate spaces for the development of new activities.

Parks created following this strategy are the result of a previous deliberated strategy that takes advantage of the positive externalities arising from a geographical concentration, an agglomeration of knowledge, and a resource-sharing policy (Phan et al., 2005).

Once the park is formalized, these parks persist in their task of incorporating innovative companies and in ensuring that technology gets improved and transferred between the university and its indoor firms. However, they also need to put an eye to the quality of their infrastructures in order to remain attractive for corporations not currently involved in the project. Thus, the commercialization function is essential in order to attract new partners and complete the clusterization puzzle.

Perhaps the main challenge that planned parks face has to do with the creation of the specific unit that will be in charge of managing the park. Because prior to the formalization of the park, the different stakeholders might have worked independently and developed their own ways of operating, the establishment of a new centralized unit might create some tensions that hamper the relationships between the stakeholders. In this respect, the development of a trusting relationship between these units and the different stakeholders is one of the main determinants of collaborative success. The higher the level of trust between partners, the greater their involvement and commitment in the alliance will be. Accordingly, building an atmosphere of trust is essential in order to contribute to the free exchange of information between committed partners. Once trust is gained, it encourages the creation of knowledge-sharing routines that enable the collaboration to be maintained over time.

As for the existence of communication networks in the area, in an unplanned park, this aspect is not as important as in the former case because transportation facilities and IT services have usually been supplied as a natural response of the spontaneous activity that emerged prior to the formalization of the park.

Stakeholders of URPs

URPs have different stakeholders. These can be broadly classified into four groups: investors (public and private), the region (e.g., represented by governments), firms and corporations, and the science-based agents (e.g., the university, research institutes, and research centres). Because interests might differ, Table 8.1 summarizes, for each stakeholder, their interests for getting involved in a URP, the advantages/difficulties associated with the creation strategy of the park (planned/unplanned), and some parameters that can be used to evaluate the potential of a URP.

TABLE 8.1 Characterization of the Two Typologies of URPs According to Stakeholders' Expectations

Stakeholders	Interest for getting involved	Planned park	Unplanned park	Parameters to evaluate the potential
Investors (public and private)	Possibility to invest in activities and firms with a high-tech component.	The risk entailed is quite high because despite the existence of the university, the interactions between university and industry have not been proved, nor has the technological potential of the region.	Low risk level as there is empirical evidence that a there is critical research mass and knowledge flows. Therefore, there is already an ongoing activity that can be further exploited.	Expected annual occupancy Number of firms requesting spaces relative to current occupancy Quality of the infrastructures Evolution of the supply/demand side Annual profitability (per building, company, park, etc.)
Region	Development of new/old urban spaces. Contribution to the re-industrialization or economic development of the area. Specialization of industrial sectors. Creation of highly skilled jobs. Increase the visibility of the region.	The park is conceived as a policy instruments to promote regional development. Doubts regarding occupancy are evidenced.	The dynamics and endogenous scientific and technological activities that already exist seem to guarantee the sustainability of the park.	Park brand and reputation of the city (both indoors and outdoors) Employment evolution Cycles of events registered Attraction of wealth

(continued)

TABLE 8.1 Characterization of the Two Typologies of URPs According to Stakeholders' Expectations (cont.)

Stakeholders	Interest for getting involved	Planned park	Unplanned park	Parameters to evaluate the potential
Firms and corporations	Access to human capital, knowledge, and innovation networks. Assistance in pre-competitive stages of product development. Use of universities' infrastructures (i.e., R&D labs) and exploitation of academics' expertise. Economic and fiscal benefits for being located in a park. Facilities for the creation of NTBF (e.g., business incubators) Outward projection.	It implies attracting companies from scratch, requiring a convincing marketing campaign, highlighting the advantages of locating there. The status and the research capacity of the university are essential. Transportation facilities, the quality of the infrastructures and IT services are critical aspects that corporation might value.	Because of the previous activity, there is some evidence of how the park would form. In order to attract new firms, the promotion strategy should be focused in highlighting the profits and benefits obtained by firms already installed.	Number of R&D executed projects Spin-offs and start-ups created Companies attracted Patents and products created Survival rates of businesses created/installed
University, research institutes/centres	Open up new research projects. Possibility of establishing synergies and knowledge exchanges with firms. Availability of high-quality support infrastructures for knowledge transfer activities Increase their reputation.	While universities might positively perceive the initiative, research centers located near other universities might show some resistance to change their location. Probably its establishment or the creation of new ones would take place after some years of operation.	From the university's viewpoint, formalizing the URP is a unique opportunity for boosting R&D activity. Despite the knowledge spillovers generated prior to the formalization of the park, the attraction of new research centers might be subject to the facilities that are offered in their current locations.	Number of research centres located Number of R&D executed projects Membership in innovation networks Advanced services supporting R&D activities Infrastructures

METHODOLOGY

Sample

The Spanish public higher education system comprises 47 universities. Because information regarding some of the variables of interest was unavailable, six universities have been dropped out from the sample: Universidad del País Vasco, Universitat de las Illes Balears, Universitat Rovira i Virgili, Universidad de Oviedo, Universidad de la Rioja, and Universidad Politécnica de Cartagena. The final database therefore comprises information about 41 Spanish universities for the period 2004–2011.

Data

As in any production system, universities are given a set of inputs and are asked to maximize their outputs. This is clearly the case in the public higher education system. In this case, the outputs refer to technology transfer achievements: number of patents granted, number of spin-offs created, and R&D contract income emerging from university–industry partnerships. For the purpose of this study, we are mainly interested in the effect that support infrastructures (i.e., URPs) have on technology transfer outputs, butwe also included as inputs human capital and financial resources, as they respectively represent the workforce and the resources needed to produce such outputs.

Proxies for the selected inputs and outputs are described below, and descriptive statistics are presented in Table 8.2. Data come from the IUNE Observatory of Spanish University Research Activity and the annual reports of the RedOTRI (Spanish Network of Technology Transfer Offices). Additional information on specific variables comes from the biannual surveys of the CRUE (Conferencia de Rectores de Universidades Españolas).

Inputs

Because one of the main contributions of this study is to investigate not only the effect of having an affiliated research park but to examine potential differences in the outputs depending on the creation strategy of the park, two dummy variables have been used to capture this effect. Table 8.3 shows the cases covered by these variables.

The classification of URPs as planned or unplanned has been conducted after a detailed website search. In most cases, websites contained information about the history and foundation of the research park, and therefore the creation strategy of the URP was easily identifiable. In other cases, the creation strategy was inferred from documents coming from different

TABLE 8.2 Descriptive Statistics for the Selected Variables

Variables			Mean	Std. Dev.	Min	Max	Observations
Inputs	Human capital	Faculty	1174.41	784.83	124.00	4,027.00	328
		TTO staff	15.19	13.08	1.00	94.00	312
	Financial resources	TTO budget (thousand euros)	693.89	698.06	1.27	3,995.00	260
	Support infrastructures	Incubator	0.56	0.50	0.00	1.00	328
		Planned park	0.21	0.41	0.00	1.00	328
		Unplanned park	0.70	0.46	0.00	1.00	328
Outputs	Technology transfer outputs	Patents	7.14	7.62	0.00	46.00	327
		Spin-offs	2.31	3.37	0.00	22.00	327
		R&D contracts income (thousand euros)	9,566.68	12,332.95	380.22	99,678.00	282

Source: Self-devised based on information from CRUE, RedOTRI, and IUNE reports.

TABLE 8.3 Variables Capturing the Creation Strategy of the Park

Case[a]	Variable		Interpretation
	Planned park	Unplanned park	
1	0	0	The univeristy does not have a URP
2	1	0	The univeristy has a URP, and it was created following a deliberate plan (planned park)
3	0	1	The univeristy has a URP, and it responds to the typology of unplanned parks

[a] The case "planned park = 1" and "unplanned park = 1" cannot take place, as an URP can only follow one of the two strategies, but not both simultaneously.

sources such as the local government, the university, or organizations related to the park. The classification list was reviewed by a panel of experts. With their input, the final classification was achieved.

Alongside URPs, many universities have also been provided with business incubators—that is, specific spaces and training services aimed at fostering the birth of new companies. While URPs tend to be large-scale projects that house everything from corporate, government, or university labs to very small companies, business incubators are specifically oriented towards helping entrepreneurs with mature or fledgling ideas that could become a business, as well as companies in a pre-competitive stage of development. This variable is proxied through a dummy variable, where 1 denotes that the university has a business incubator and 0 the absence.

Two additional dimensions are included in the analysis: human capital and financial resources. As for the former, human capital represents the intellectual capital that universities use to attain their objectives. Consequently, human capital, understood in terms of knowledge, abilities, and capabilities provided by individuals, constitutes a key determinant to ensuring corporate success. For the purposes of this study, human capital is operationalized using two indicators: (a) faculty members, representing the workforce in charge of conducting research activities that result in technology transfer outcomes, and (b) the personnel working in the technology transfer office that provide guidance and specific support to technology transfer activities.

The second dimension refers to financial resources. This economic cushion is what allows universities to access new resources and materials to develop new activities, which might result in new developments, prototypes, and engines. Financial resources are particularly relevant in engineering and health-related disciplines where research projects tend to be linked with the acquisition of new machinery or raw materials that usually require high investments (Landry, Amara, & Ouimet, 2007). In this study, we have considered the budget of the TTO as a proxy for the financial resources.

Outputs

The first output considers the number of patents granted. Patents are a way of safeguarding technological discoveries. Although there is no guarantee that a patent will be exploited, patents represent a unique asset with a high potential for further market applications. A university highly committed with technology transfer activities will encourage its researchers to engage in valorization activities and, thus, will exhibit a higher number of patents granted.

The second output is measured through the number of spin-offs. University spin-offs result from an entrepreneurial process initiated in a university setting and are based on the exploitation of a university development (Rasmussen, 2008). Spin-offs have an important role outside universities' doors. They are seen as mechanisms able to regenerate the economic activity within the region through the exploitation of the most up-to-dated technology available at universities.

Lastly, the third output accounts for university–industry partnerships. We use the income from R&D contracts through which enterprises and universities collaborate on a win–win basis. Because the third mission of universities aims at narrowing the gap between science and industry, we consider of great importance including a variable that captures the economic activity that these partnerships generate.

Regression

In order to assess the effect of URPs on university technology transfer outputs, three models are tested, one for each of the technology transfer outputs listed above. Models 1 and 2 refer to the activities of patenting and creating new academic ventures, respectively. In both cases, a binomial regression analysis is chosen, as this method best fits to the characteristics of the dependent variables (count variables exhibiting a highly skewed distribution) (Greene, 2003, 2008). As for the establishment of R&D contracts (Model 3), the method chosen is linear regression.

To ensure the robustness of the significance of the results, we also examined the explanatory power that each exogenous variable has over the corresponding research output in an individual fashion. Results validate the consistency of the three models.

To obtain normality some variables were transformed by use of the natural logarithm (faculty, TTO staff, TTO budget, and R&D income). Normal probability plots of the residuals for Model 3 corroborate that they were normally distributed.

Cluster Analysis

Cluster analysis is a technique that allows identifying groups of observations with different behavioral paths, given the presence of specific variables that are expected to influence the sampled units (Everitt, 1980). For the purpose of this research, we use a nonhierarchical cluster analysis (*k*-means), employing the inputs and outputs detailed earlier.

The cluster analysis is based on the Euclidean distance between vectors of the standardized values of the variables under analysis (Anderberg, 1973; Everitt, 1980). Through this procedure, observations are classified according to the similarities of organizational and environmental dimensions. The K-means cluster analysis requires the establishment of a fixed number of clusters. This represents the main pitfall of nonhierarchical cluster analysis, because in many research fields (including social sciences), cluster analyses are often exploratory. This chapter adopts two approaches to corroborating the number of clusters and the validity of the analysis. First, the study proposes the Calinski and Harabasz (1974) statistic. This index is obtained as

$$CH(K) = \frac{B(k)/(k-1)}{W(k)/(n-k')},$$

where $B(k)$ and $W(k)$ are the between and within-cluster sums of squares, with k clusters. Since the between-cluster difference should be high and the within-cluster difference should be low, the largest $CH(k)$ value indicates the best clustering. The number of clusters that maximizes the $CH(k)$ index is 3 (*pseudo-F value:* 277.08). Therefore, the final nonhierarchical cluster asks for a three-way division. Second, a discriminant analysis further validates the cluster analysis. The results from the discriminant analysis (Table 8.4) indicate that the approach is appropriate.

TABLE 8.4 Results of the Discriminant Analysis

	Classification			
True Groups	**1**	**2**	**3**	**Observations**
Group 1	37 (100.00%)	0 (0.00%)	0 (0.00%)	37
Group 2	1 (2.33%)	40 (93.02%)	2 (4.65%)	43
Group 3	1 (0.56%)	3 (1.69%)	174 (97.75%)	178
Total	**39 (15.12%)**	**43 (16.67%)**	**176 (68.22%)**	**258**

RESULTS AND DISCUSSION

Determinants of Technology Transfer Outputs

Results from the three regression models are displayed in Table 8.5. The overreaching conclusion is that having a URP seems to play a role in all three outputs.

As for the potential differences due to the creation strategy, our results suggest that when examining the number of patents both planned and unplanned parks help in increasing the outputs, although the contribution (beta value) of the unplanned parks variable in explaining the output is higher.

Differences are, however, found when the outputs to be produced are a new academic venture or when considering the income from R&D

TABLE 8.5 Regression Results

	Model 1: Patents	Model 2: Spin-offs	Model 3: R&D contracts
Faculty	0.77*** (0.08)	0.41*** (0.12)	0.65*** (0.07)
TTO staff	0.16** (0.08)	0.06 (0.11)	0.29*** (0.09)
Budget TTO	0.13** (0.06)	0.04 (0.07)	0.08 (0.07)
Business incubator	–0.18** (0.09)	0.17 (0.14)	0.09 (0.09)
Planned park	0.69*** (0.18)	0.09 (0.21)	0.20 (0.16)
Unplanned park	0.84*** (0.21)	0.81*** (0.24)	0.62*** (0.19)
Intercept	–5.22*** (0.57)	–2.88*** (0.84)	2.61*** (0.50)
F-test (degrees of freedom, *N*)			48.46*** (6, 228)
R^2			0.58
RMSE			0.67
Log likelihood	–711.33	–496.57	
PseudoR^2 (degrees of freedom)	0.11 (6)	0.07 (6)	
Wald χ^2 (degrees of freedom)	241.38*** (6)	86.61*** (6)	
Observations	257	257	235

Note: Robust standard errors adjusted by heteroskedasticity are presented in parenthases.
*, **, *** indicate significance at the 10%, 5%, and 1%, respectively.

contracts. Those parks that result from the formalization of an endogenous and spontaneous logic are more prone to create spin-offs and establish university–industry partnerships that result in fruitful R&D contracts. On the contrary, having a planned park and not having a park seems to be indifferent. This result signals that creating and consolidating an entrepreneurial culture is something that cannot be easily achieved in an environment with no previous tradition and where routines have already been pre-established.

Concerning the other inputs included in the models, results confirm that direct labor force, namely researchers, are fundamental inputs. As for the administrative and technical staff working at TTOs, results corroborate previous studies (Berbegal-Mirabent, Sabaté, & Cañabate, 2012) indicating that they facilitate the administrative procedures linked to the processes of protecting and safeguarding invention disclosures (Model 1) and also perform a paramount role as knowledge brokers between science and industry (Model 3).

A positive and statistical significance is found for financial resources in Model 1. Following an earlier discussion, the access to funding allows researchers to work with better materials and equipment. This effect, however, dilutes in Models 2 and 3, indicating that perhaps the skills and the expertise of the technical staff rather than the access to financial resources or the size of the technology transfer office is what really enhances spin-off creation and the singing of university–industry alliances for the provision of R&D services (Berbegal-Mirabent, García, & Ribeiro-Soriano, 2015).

The effect of the business incubator also deserves a special attention. While in Models 2 and 3 such an advanced infrastructure does not seem to exert any influence, a negative sign is observed in Model 1. The underlying rationale behind this result may lie in the fact that academics have a strong preference for safeguarding their knowledge and only think about getting involved in entrepreneurial activities if the patent they applied for is granted. Therefore, business incubators might be seen as a complementary service that instead of facilitating the creation of a unique and cutting-edge new device is biased towards the promotion of new business creation.

Cluster Analysis

Because of some missing information, not all universities have been clustered for all years. Only in 6 out of the 8 years under analysis, data were complete for all clustering variables. From the cluster analysis, a consistent categorization of universities is obtained (see Table 8.6) into three different clusters. Results from this analysis show that universities tend to remain in the same cluster over time. In only five cases (UC3M, UPC, UPM, UPV, and USC) did universities move from one group to another. However, once

TABLE 8.6 Classification of Universities Based on the Cluster Analysis

Cluster	Observations	Universities
#1	6	Barcelona (UB), Carlos III (UC3M), Politècnica de Catalunya (UPC), Politécnica de Madrid (UPM), Politècnica de València (UPV), Santiago de Compostela (USC)
#2	5	Complutense (UCM), Granada (UGR), Zaragoza (UNIZAR), Sevilla (US), València (UV)
#3	30	Alicante (UA), Autònoma de Barcelona (UAB), Alcalá de Henares (UAH), Almería (UAL), Autónoma de Madrid (UAM), Burgos (UBU), Cádiz (UCA), Castilla-La Mancha (UCLM), Córdoba (UCO), La Coruña (UDC), Girona (UDG), Lleida (UDL), Huelva (UHU), Jaén (UJA), Jaume I (UJI), Laguna (ULL), Las Palmas de Gran Canaria (ULPGC), Murcia (UM), Málaga (UMA), Miguel Hernández (UMH), Pública de Navarra (UNAVARRA), Extremadura (UNEX), Cantabria (UNICAN), León (UNILEON), Pompeu Fabra (UPF), Pablo Olavide (UPO), Rey Juan Carlos (URJC), Salamanca (USAL), Valladolid (UVA), Vigo (UVIGO)

the change occurs, the position in the new cluster remains unchanged in the subsequent years. In these cases, universities are classified based on the most recent group where they are clustered.

Cluster 1 comprises three polytechnic universities (UPC, UPM, and UPV), clearly focused on engineering degrees and traditionally more involved in commercializing research result to the marketplace. They are entrepreneurial universities in that the formalization of the park results from the need of organizing and providing the appropriate spaces that support the valorization of research results. UPC, UPM, and UPV share the characteristic of having access to human and financial resources, together with the existence of a favorable environmental setting that allows them to effectible translate research results to the marketplace. Indeed, these universities have focused their efforts on a unique direction, pledging their resources and commitment to the creation of knowledge with potential commercial applications (patents) and its subsequent valorization in the marketplace (spin-offs). Also, university–industry partnerships (measured through R&D contracts income) result in lucrative agreements for the university.

While the three polytechnic universities follow a similar pattern, it is worth analysing the particularities of the three remaining universities of this group. UB bases its strategy in disseminating the knowledge stock through scientific publications. Their faculty members are really good at doing so, but they leave aside the possibility to translate the research results into something valued by the market (e.g., patents, spin-offs, R&D contracts). USC goes a step further, and besides being highly active in publishing in academic journals, it tries to take advantage of the natural spill-overs that arise from the adoption of an entrepreneurial culture. Despite its research, outputs in terms of patents and spin-offs are still a bit far from that of polytechnic universities; in recent years, there has been a substantial growth in the number of faculty members involved in technology transfer objectives. As a result, the number of patents has increased. Somehow different is the case of UC3M, a small university in terms of size (number of faculty members and students) but with a high investment in resources, evidencing the importance that technology transfer has acquired within the objective function of this university. This abundant provision of resources facilitates the establishment of university–industry partnerships for the development of R&D activities and their exploitation. Nevertheless, the patenting and spin-off activity is still in its infancy.

All universities in this group already have a URP in 2004 (with the exception of UPC, formalized in 2005); consequently, we cannot directly evaluate the impact that the formalization/creation of the park has had on technology transfer outputs. Nevertheless, from the first stage analysis (regression analysis), we can conclude that planned parks (the creation strategy followed by almost all of the universities in this group) help explain why these universities outperform in all the outputs under analysis.

Cluster 2 comprises those universities with the greatest critical mass of faculty members, but they have a limited number of TTO staff supporting research activities. As for budget, they are ranked second (Table 8.7),

TABLE 8.7 Average Descriptive Statistics for the Cluster Analysis

Variables		Cluster 1	Cluster 2	Cluster 3
Inputs	Faculty	1,741.11 (727.08)	2,532.79 (701.22)	837.72 (396.50)
	TTO staff	40.28 (17.94)	14.99 (5.88)	11.86 (8.33)
	TTO budget	2,182.69 (620.10)	649.91 (244.80)	394.71 (258.35)
	Business incubator	0.73 (0.45)	0.49 (0.51)	0.57 (0.50)
	Planned park	0.32 (0.47)	0.30 (0.46)	0.81 (0.39)
	Unplanned park	0.65 (0.48)	0.65 (0.48)	0.04 (0.21)
Outputs	Patents	15.97 (13.00)	13.53 (6.66)	4.67 (4.19)
	Spin-offs	4.38 (4.84)	3.95 (3.50)	1.66 (2.08)
	R&D contracts income	21,506.03 (18,531.64)	17,846.48 (13,886.69)	5,243.73 (4,041.44)

Note: Standard deviation in parentheses.

but quite far from that of universities in cluster 1 (in average, 70% less). Although all of them had a research park in the years surveyed, only half of them (55%) had a business incubator. In absolute terms, universities in this group are very active in transferring research results to the marketplace through spin-offs and patents. Also, the income from R&D contracts is remarkable. However, when considering the ratios of productivity (dividing by the number of faculty), they their results are not that good, particularly in terms of spin-offs.

All in all, this path suggests that although these universities have the potential and possess the resources to transform knowledge into marketable results, only a small proportion of their faculty members is engaged in technology transfer activities. Incentives to actively participate in technology transfer activities seem therefore to be rather low or inefficient. However, it is worth pointing out that those universities with a planned research park outperform their counterparts, particularly in terms of spin-offs, validating the results obtained in the previous stage.

Finally, cluster 3 reveals that universities in this group lack internal resources and capabilities. This resource shortage is observed in terms of human capital. They also have a restricted financial capacity and the availability of appropriate research infrastructures for the development of their daily activities is limited. Moreover, universities in this group are characterized by their limited size. As for technology transfer outputs, these universities do not seem to excel at in any of them compared to other clusters. Notwithstanding, it appears that they are more comfortable disseminating research through publications than through more market-oriented outputs.

An in-depth analysis of the universities in this group indicates that in most instances (81%), the creation strategy of the park follows a planned scheme. It is noteworthy to highlight that until recently, some of the universities included in this group had not yet formalized/created the research park. Because the effect of the park cannot be immediately perceived in terms of the outputs produced, the interpretation of the effect of the park might be taken with a grain of salt for certain universities. Nevertheless, there are some remarkable cases that deserve further attention. For instance, since the formalization of the URP of the UAB, the number of patents and R&D contracts income has substantially increased. ULL also exhibits a similar behavior. Prior to the creation of the park, this university had not issued any patent and revenues coming from R&D collaborations with firms were limited. Since the establishment of the park, these activities are increasingly growing. Figures from UVA also follow a similar evolution. UVA's URP boosted spin-off creation and helped to more than double the conversion rate of invention disclosures into patents. Similar positive effects played by the URP are observed in UDL, UHU, UJI, UMH, UNEX, and URJC. While research outputs of these universities have not suffered from significant changes in the recent years, since the creation of the park, the commercialization of research results has improved.

To sum up, universities from this group exemplify to some extent the positive effect of the URP, reinforcing the argument that URPs act as catalysts for the development of technology transfer activities.

DISCUSSION

Conclusions

Among the different typologies of science parks, it is possible to distinguish those parks that are closely related to a university. This category, namely university research parks, fulfills the specific task of adding well-equipped infrastructures for academic R&D, but it also demands a highly structured strategy to facilitate research commercialization. URPs are expected to strengthen the universities' capacities and help them to accomplish their missions, particularly in their contribution to regional development.

However, the creation of URPs is a complex phenomenon that requires special attention. Although there are many studies and reports examining URPs, only a few of them are focused on the first stages of the establishment of a park. In this study we put forward that an analysis of the strategies that were followed and the mechanisms that drove the creation of URPs deserve further attention.

In this study, we first embarked on an exhaustive literature review. Nevertheless, we found that literature is scarce. Based on the evidence, we posit that two opposite strategies exist depending on the desire of starting and building something from scratch or basing on a previous spontaneously innovative activity that pushes the formalization of a park. Accordingly, we distinguish *planned parks* from *unplanned* ones. While the former respond to an internal design and a planned structure, in the latter, the desire to establish a park results from an organizational need to align in a systematic way the sum of individual activities that have emerged around the university. Both strategies have been presented and their implications have been discussed.

Next, we conducted an empirical analysis based on the Spanish public higher education system. Results suggest that in the case of patents, the existence of a research park does have a strong impact on the number of patents granted; however, the creation strategy of the URP does not seem to affect the number of patents granted. In contrast, for spin-offs and R&D contracts, the creation strategy of the park is a paramount factor. Not only is the presence of the park necessary, but it is important to have developed an entrepreneurial culture. R&D contracts and spin-offs represent a more direct route to transfer research knowledge to the industry, and thus, the involvement of all stakeholders is more critical than in the case of patents. This entrepreneurial culture is usually more common in the case of unplanned parks, as the relationship between university and industry is already in place. In the case of planned parks, developing the necessary culture to foster R&D contracts and spin-offs takes more time and effort.

Implications for Policy

From our analysis, some policy lessons can be drawn:

- Universities represent the cornerstone of research activities, education, and innovation. Accordingly, they are important vehicles for regional development and are a key component for the economic regeneration and growth of territories.
- Academic entrepreneurship is a crucial component necessary to consolidate knowledge transfer processes and contribute to renew the local industry fabric.
- Articulating a URP (which includes R&D, engineering, production, marketing, and maintenance, integrating diverse skills) is a difficult task due to the variety of agents involved with different objectives, interests and ways of operating.
- Innovation and technology development is not just a matter of URPs. URPs are important instruments that facilitate them, but not the unique existing ones. In that respect, our study evidences that the creation strategy of the park, whether planned or unplanned, does make a difference in the output of the university.

Directions for Future Research

Further research endeavors should now be directed towards improving the understanding of the implications that entail following each of the two creation strategies described (*planned* and *unplanned*). Because of the aforementioned lack of information, studies addressing this issue should be probably conducted under a case study focus.

Limitations

An important limitation of this study refers to the lack of a comprehensive database with specific indicators of Spanish URPs (e.g., number of firms located in the park, economic performance, etc.). Also, it would have been interesting to have access to data before 2004, as many Spanish URPs started operating in the late 1990s.

ACKNOWLEDGMENTS

The authors would like to thank Dr. Francesc Solé Parellada (Universitat Politècnica de Catalunya) for his insightful comments during the conceptualization and preparation of the manuscript.

REFERENCES

Anderberg, M. R. (1973). *Cluster analysis for applications.* New York, NY: Academic Press.

Berbegal-Mirabent, J., García, J. L. S., & Ribeiro-Soriano, D. E. (2015). University–industry partnerships for the provision of R&D services. *Journal of Business Research, 68*(7), 1407–1413.

Berbegal-Mirabent, J., Martín, M. E., & Solé, F. (2010). Los parques universitarios: Gestión del conocimiento, espíritu innovador y dinamización del entorno local. El caso del Parc de Recerca i Innovació de la Universitat Politècnica de Catalunya. In P. Díaz de Basurto, E. Cilleruelo, C. Ochoa, P. Ruiz de Arbulo, E. Zarrabeitia, I . Alvarez & I. Heras (Eds.), *Book of Full Papers from the 4th International Conference on Industrial Engineering and Management: Industrial Engineering as University Third Mission agent* (pp. 1977–1987). San Sebastián, Spain: Department of Management of the Universidad del País Vasco.

Berbegal-Mirabent, J., Sabaté, F., & Cañabate, A. (2012). Brokering knowledge from universities to the marketplace: The role of knowledge transfer offices. *Management Decision, 50*(7), 1285–1307.

Bigliardi, B., Dormio, A. I., Nosella, A., & Petroni, G. (2006). Assessing science parks' performances: Directions from selected Italian case studies. *Technovation, 26*(4), 489–505.

Braczyk, H., & Cooke, P. H. (1998). *Regional innovation systems: The role of governances in a globalized world.* London, England: University College London.

Calinski, R. B., & Harabasz, J. (1974). A dendrite method for cluster analysis. *Communications in Statistics, 3*(1), 1–27.

Chen, K., & Kenney, M. (2007). Universities/research institutes and regional innovation systems: The cases of Beijing and Shenzhen. *World Development, 35*(6), 1056–1074.

Chorda, M. (1996). Towards the maturity stage: An insight into the performance of French technopoles. *Technovation, 16*(3), 143–151.

Clark, B. R. (1998). *Creating entrepreneurial universities: Organizational pathways of transformation.* Oxford, England: Pergamon.

Cooke, P. H., & Morgan, K. (1994). The creative milieu: A regional perspective on innovation. In M. Dodgson & R. Rothwell (Eds.), *The handbook of industrial innovation* (pp. 25–32). Cheltenham, England: Edward Elgar.

CRUE. (2006). *La universidad española en cifras. Información académica, productiva y financiera de las universidades españolas. Año 2004. Indicadores Universitarios.* J. Hernández-Armenteros (Ed.). Madrid, Spain: Conferencia de Rectores de las Universidades Españolas.

CRUE. (2008). *La universidad española en cifras. Información académica, productiva y financiera de las universidades españolas. Año 2006. Indicadores Universitarios.* J. Hernández-Armenteros (Ed.). Madrid, Spain: Conferencia de Rectores de las Universidades Españolas.

CRUE. (2010). *La universidad española en cifras. Información académica, productiva y financiera de las universidades españolas. Año 2008. Indicadores Universitarios.* J. Hernández-Armenteros (Ed.). Madrid, Spain: Conferencia de Rectores de las Universidades Españolas.

Etzkowitz, H., Webster, A., Gebhardt, C., & Cantisano Terra, B. R. (2000). The future of the university and the university of the future: Evolution of ivory tower to entrepreneurial paradigm. *Research Policy, 29*(2), 313–330.

Everitt, B. S. (1980). *Cluster analysis* (2nd ed.). London, England: Heineman.

Greene, W. (2003). *Econometric analysis* (5th ed.). Upper Sadler River, NJ: Prentice-Hall.

Greene, W. (2008). Functional forms for the negative binomial model for count data. *Economics Letters, 99,* 585–590.

Hansson, F. (2007). Science parks as knowledge organizations: The "ba" in action? *European Journal of Innovation Management, 10*(3), 348–366.

IUNE. (n.d.). *IUNE observatory.* Retrieved from http://www.iune.es/

Koh, F. C. C., Koh, W. T. H., & Tschang, F. T. (2005). An analytical framework for science parks and technology districts with an application to Singapore. *Journal of Business Venturing, 20*(2), 217–239.

Landry, R., Amara, N., & Ouimet, M. (2007). Determinants of knowledge transfer: Evidence from Canadian university researchers in natural sciences and engineering. *Journal of Technology Transfer, 32*(6), 561–592.

Leyden, D. P., Link, A. N., & Siegel, D. S. (2008). A theoretical and empirical analysis of the decision to locate on a University Research Park. *IEEE Transactions on Engineering Management, 55*(1), 23–28.

Leydesdorff, L., & Etzkowitz, H. (1996). Emergence of a triple helix of university–industry–government relations. *Science and Public Policy, 23*(5), 279–286.

Link, A. N., & Scott, J. T. (2003). U.S. science parks: the diffusion of an innovation and its effects on the academic missions of universities. *International Journal of Industrial Organization, 21*(9), 1323–1356.

Link, A. N., & Scott, J. T. (2006). US university research parks. *Journal of Productive Analysis, 25*(1), 43–55.

Link, A. N., & Scott, J. T. (2007). The economics of university research parks. *Oxford Review of Economic Policy, 23*(4), 661–674.

Löfsten, H., & Lindelöf, P. (2001). Science parks in Sweden. Industrial renewal and development. *R&D Management, 31*(3), 309–322.

Löfsten, H., & Lindelöf, P. (2002). Growth, management and financing of new technology-based firms: Assessing value-added contributions of firms located on and off science parks. *Omega, 30*(3), 143–154.

Markman, G. D., Phan, P. H., Balkin, D. B., & Gianiodis, P. T. (2005). Entrepreneurship and university-based technology transfer. *Journal of Business Venturing, 20*(2), 241–263.

MacDonald, S. (1987). British science parks: Reflections on politics of high technology. *R&D Management, 17*(1), 25–37.

Phan, P., Siegel, D. S., & Wright, M. (2005). Science parks and incubators: Observations, synthesis and future research. *Journal of Business Venturing, 20*(2), 165–182.

Rasmussen, E. (2008). *Spin-off venture creation in a university context. An entrepreneurial process view.* Bodø: Bodø Graduate School of Business.

RedOTRI. (2004). *Informe de la encuesta RedOTRI 2004.* Madrid, Spain: RedOTRI–Universidades (CRUE).

RedOTRI. (2005). *Informe de la encuesta RedOTRI 2005.* Madrid, Spain: RedOTRI–Universidades (CRUE).

RedOTRI. (2006). *Informe de la encuesta RedOTRI 2006.* Madrid, Spain: RedOTRI–Universidades (CRUE).

RedOTRI. (2007). *Informe de la encuesta RedOTRI 2007.* Madrid, Spain: RedOTRI–Universidades (CRUE).

RedOTRI. (2008). *Informe de la encuesta RedOTRI 2008.* Madrid, Spain: RedOTRI–Universidades (CRUE).

RedOTRI. (2009). *Informe de la encuesta RedOTRI 2009.* Madrid, Spain: RedOTRI–Universidades (CRUE).

RedOTRI. (2010). *Informe de la Encuesta de Investigación y Transferencia de Conocimiento 2010 de las Universidades Españolas.* Madrid, Spain: RedOTRI–Universidades (CRUE).

RedOTRI. (2011). *Informe de la Encuesta de Investigación y Transferencia de Conocimiento 2011 de las Universidades Españolas.* Madrid, Spain: RedOTRI–Universidades (CRUE).

Siegel, D. S., Westhead, P., & Wright, M. (2003). Assessing the impact of university science parks on research productivity: Exploratory firm-level evidence from the United Kingdom. *International Journal of Industrial Organization, 21*(9), 1357–1369.

Squicciarini, M. (2008). Science Parks' tenants versus out-of-park firms: Who innovates more? A duration model. *Journal of Technology Transfer, 33*(1), 45–71.

CHAPTER 9

COTTON TECHNOLOGY INTRODUCTION AND CROP IMPROVEMENT IN SHANXI DURING THE REPUBLIC OF CHINA PERIOD

Na Liang and Tao Shi
Shanxi University of Finance and Economics, Shanxi, China

Fei Wang
Shanghai Academy of Social Sciences, Shanghai, China

ABSTRACT

Using historical ethnography methodology, this chapter presents a detailed analysis of governmental measures and outcomes in the introduction of cotton technology and crop improvement by Shanxi provincial government by citing a large amount of original historical data and explains the policy support for cotton industry expansion and technological progress in Shanxi Province during the period 1911–1945 ROC government period.

Since the Southern Song Dynasty, the production technology of intensive agriculture farming has developed to an extent that small investment has

Technological Innovation Networks, pages 193–210

been unable to achieve innovation. In the Ming and Qing Dynasties, agriculture was able to satisfy a growing population, thanks to the introduction of high-yield, cold- and drought-resistant crops such as potatoes and corn from South America. The planting area of these two crops had constantly expanded in Shanxi since the Qing Dynasty, and they replaced wheat to become the main food crops in northern China. Moreover, handicraft technological advances were also limited. In the absence of sufficient capital and capacity for internal technological innovation, importation of technical equipment became the main means to improve technology and production output in Shanxi and even in the whole of China. Global cotton demand surged during the period after the founding of Republic of China, and the ROC government thus vigorously advocated cotton planting. Shanxi also considered cotton planting as an important part of its "10-Year Construction Program" and introduced and improved high-quality cotton varieties by setting up experimental cotton farms and supported indemnified experimental farms to actively promote development of the cotton industry.

In the research into development of the cotton industry in the Republic era, *Development of China's Cotton Industry* by Zhongping Yan (1943) is the earliest and most systematic study of the development of the cotton industry during the Republic of China era. By conducting a survey of 536 counties (districts) in 16 provinces, we carried out an in-depth investigation into the development, progress, and distribution of the handmade cotton textile industry in the 1930s and analyzed the background conditions of China's development of industrial capitalism (Yan, 1943). On this basis, Lin Gang supplemented the data on the handmade cotton industry in 17 cities of China in 1930s and proposed that the number of counties growing cotton collectively made one fifth of the cotton industry of Shanxi Province, but total production was far from meeting internal demands in the province, which still mainly depended on imports (Lin, 2002). The cotton textile industry was one of the pillar industries of the Republic of China, and Shuijin He (2001) believed that it was influenced not only by international events, but it was affected much more by the factors of supply and demand. China experienced many occasions of boom and bust in its textile industry during 1914–1925. The cause of the boom times was market disequilibrium in raw materials and products, while the reason for the bust was unchecked growth and expansion after high profits were realized (He, 2001, pp. 167–175). Such cycles did not disrupt development of the textile industry, but on the contrary, it promoted technical progress. Shandong as a major cotton producing province, Weimin Zhuang (1996) mentioned in "Introduction of New Varieties of Crops by Shandong in Modern Times and Its Impacts" that the modernization of agricultural production led to improvement in agricultural productivity, and the introduction and promotion of crops had the greatest impact on the agricultural economy and production. *Development of*

Shandong's Commodity Economy in Ming and Qing Dynasties by Tan Xu (1998) explored the development of commodity economy in Shandong during Ming and Qing Dynasties using commodity trading statistics. The Yangtze River Delta region as a center of technical post-processing also aroused the attention of scholars; examples can be found in discourse expositions including *Overview of Modern Cotton Industry in Jiangsu Province* (Zhang,1991), *Study of Foreign Trade of Cotton in Yangtze River Delta during 1912–1936* (Yu, 2010), and *Research on Shanghai's Cotton Planting History* (Wang,1984).

The time span of studies conducted by international scholars on China's textile industry extend to even earlier periods, such as *Handicraft and Manufactured Cotton Textiles in China 1871–1910* (Feuerwerker, 1970). In *China's Silk Trade: Traditional Industry in the Modern World* (Lillian, 1981), the silk industry as a part of the cotton and textile industry is encouraged by Chinese government policies and driven by commerce development in the international market place, which reflected the significance of its exports to the Chinese economy as a whole. Foreign scholars also showed interest in the development of the cotton textile industry in China during the Republic era with writings including *Merchants, Peasants and Industry: the Political Economy of Cotton Textiles, Nantong County, 1895–1935* (Walker, 1986), which covers a special on-site investigation and data collection in Nantong County, documenting businessmen and workers engaged in the cotton and textile industry in Nantong County.

As an emerging success story of the cotton and textile industry, Shanxi has been the center of government attention. "Implementation and Performance of Shanxi's Cotton Policies During 1917–1937" (Xu, 2012) studied the active cotton policy implemented by the ROC government, which to some extent resolved the problems of the cotton industry at that time. "Development of Railway and Modern Cotton Industry of Shanxi: 1907–1937" (Li, 2010) focused on an analysis of the impacts of rail transport on the development of cotton commercialization. During WWII, the Japanese government also attached great importance to the development of the cotton textile industry in Shanxi. *The Overview of Main Workshop in Shanxi* made a detailed analysis of six large textile factories in Shanxi (NO. 3500 department [Japan], 1942). It contains output statistics after introduced of improved cotton varieties. So far, the research studies by both domestic and foreign scholars mainly focused on the impacts on production and trade of cotton on the modernization of Shanxi; however, there were no systematic studies on the technological advances and upgrades in Shanxi's cotton industry. This chapter aims to give a detailed introduction to the development history of Shanxi's cotton industry by looking at the establishment of cotton industry research institutions, improvement and widespread adoption of cotton varieties selection techniques, and improvements in cotton technology, harvest yield, and quality.

ESTABLISHMENT OF COTTON RESEARCH INSTITUTES

From the late Qing Dynasty, the Western production model using large machinery was introduced to China. Because of high profits, light industries such as silk reeling and textile spinning became the top choices for government-run or privately run enterprises. Suddenly, demands for cotton soared. During the late Qing Dynasty period, the imperial provincial governor of Shanxi Hu Pingzhi tried to build a fully mechanized cotton mill in Jiang County of Shanxi Province; however, deficient transport infrastructures led to "large machinery couldn't be taken into Shanxi which revealed flaws in the plan. As a result, the construction of the cotton mill came to an end before it was half built. Purchased machine parts ended up staying in around in various customs ports in Shanghai and Hankou" (Editor of *Jinyang Daily*, 1936, p. 71, author translation). As a matter of fact, Shanxi saw low cotton output in the late Qing Dynasty period. "Even though a few were planting cotton along the Hedong Corridor, but as the peasants were illiterate at that time, with a lack of transport facilities, hence, the meager output was only enough for daily consumption with no surplus stock for sales" (Industrial International Trade Bureau, 1937, p. 467, author translation). Therefore, in order to develop the textile industry that could earn good profits back then, local cotton output needed to be raised.

In ROC Year-6 (1917), Shanxi Province set up a cotton experimental farm in Lingfeng Prefecture of Linfen and designated a Shanxi Agricultural College graduate, Chen Jian, to take charge of administration to introduce cotton varieties from the U.S. and Germany for local farmers to grow. He also compiled the cotton planting methods in colloquial language to instruct cotton farmers on how to take care of cotton fields from eight aspects: namely, field preparation, soil selection, seed selection, seed cultivation, seedbed preparation, transplantation, seedling spacing, weeding tillage, topping, and fertilizing (Editor of Industrial News Highlights, 1918). As the cotton experimental farm was directly administered under the governor's office, managers directly reported to the governor, Yan Xishan. His aims of the establishment roughly came down to: first, to carry out experiments in various aspects regarding cotton plantation such as climate, soil and fertilizers, and investigate field conditions in different locations to select locations most suitable for cultivating different varieties of cotton; second, to identify the superior varieties by determining fiber qualities and recommend varieties that should be used in planting fields of different locations; and third, to improve cotton plantation technologies and guide cotton farmers in the production of cotton (Historical Annals Research Institute of Shanxi Province, 2001).

In November 1917, the government of Shanxi Province promulgated the rules for cotton experimental farms, which provided cotton experimental farms should have one manager, two technicians, and two clerks to be responsible for guiding seed selection and distributing seeds to cotton farmers for free. During cultivation and harvest, staff shall tour the plantations to give on-site guidance, answer questions raised by cotton farmers in detail, deal with all sorts of matters including preventing insect infestation, and report the harvest to the provincial governor's office. A liaison coordinator would be chosen mainly from farmers enthusiastic about cotton planting, with no restriction on the number of people to trial such technologies as cultivation, reaping, and fertilization. In addition, during the slack season, technicians from the government-run cotton experimental farm would conduct training courses in counties to teach cotton planting technologies. From ROC Year-10, further experimental cotton farms were successively set up in other counties including Taigu, Wenshui, Gaoping, Dingxiang, Hai County, and Linfen, and thereafter cotton plantation area underwent continuous expansion.

In 1928, due to various agricultural needs, the government of Shanxi Province transformed the experimental cotton farms into agricultural experimental agriculture and cotton farms, while still retaining the responsibility of encouraging cotton plantation and cottonseed cultivation. The year 1932 saw the experimental agriculture and cotton farms reverted back to experimental cotton farms charged with the special responsibility of promoting cotton plantation. The provincial government also formulated the Cotton Industry Special Project, with the promotion of cotton industry as its primary task, while aiming to improve quality by cultivating high-quality varieties (Statistical Bureau, 1935).

In 1933, a main experimental cotton plantation and other branch plantations were established in Hedong and Yanning Districts under the administration of the Industry Department. The main experimental cotton plantation was located in Linfen with branch plantations in the Yuci region, dividing the province into ten districts with five districts in Linfen, Quwo, Wanquan, Hai County and Xi County in Hedong District covering 35 counties in total; a further five district were in Yuci, Wenshui, Qin County, Gaoping and Guo County under the administrative prefecture of Yanning, covering a total of 47 counties. Each district in Hedong was assigned an instructor while each district in Yanning was assigned two instructors to provide field guidance and supervise the farmers to make improvements.

The principal business and organization in the main experimental cotton plantation and branch plantations in Shanxi Province focused on research on cotton cultivation technologies and promotion of cotton cultivation, ranging from studying the biophysical features of cotton, cultivating fine varieties for specific local soil conditions, preventing and treating insect infestation, and improving new farm implements. These entities played a critical role in promoting cotton in Shanxi Province during the Republic era.

TABLE 9.1 Summary of Organizational Establishment, Functions, Salaries and Expenditures in the Main Experimental Cotton Plantation and Branch Plantations in Shanxi Province

	Organization	Staff		Duties	Salary (Yuan/head)	Expenditures (Yuan/head)
		Post	No.			
Main plantation		Manager	1	Take overall responsibility of running the farm	90	
	Administrative section	Administrative director	1	Routine work including receiving and dispatching correspondence, filing documents and safekeeping of seals, compiling statistics, accounting, storage and sales of cotton products.	90	280
		Clerk	1–2		30	
	Technical ection	Technical director	1	Technical affairs such as breeding, cultivation experiments, biophysical research on cotton, insect infestation prevention and treatment, quality variety cultivation, editing science popularization literature, perfecting new farming implements and collecting specimen.	90	1680
		Clerk	1–2		40	
	Promotion ection	Head of promotion	1	Promotional matters including training technical instructors, dissemination of cotton planting technologies, guiding cotton cultivation, compiling estimates and statistics of cotton output, grading cotton, exploring sales channels, dividing and surveying cotton planting regions, giving instructions on prevention and treatment of insect infestation, promoting, purchasing and distributing high-quality seeds, guiding cotton ginning and answering queries.	90	
		Instructor	5		24	
	Total					5440
Branch		Manager	1	Same duties as similar organization units in the main plantation.	60	
		Clerk	1		30	
		Technician	1		60	
		Instructor	5		24	

Source: Agriculture Department, 1935a

In 1934, Shanxi established Cotton Production Improvement Stations, designating volunteer cotton production surveyors in all villages with responsibility for surveying farm produce and cotton output in every village and reporting to the Cotton Production Improvement Stations six times a year, with the initial plot estimate report to be submitted before July 5th, a winter crops total harvest area and harvest volume estimate report to be submitted before August 5th, a second plot estimate report to be submitted before September 5th, a harvest volume survey special report to be submitted before October 5th, a third plot estimate report to be submitted before November 5th, and a summer crops total harvest area and harvest volume estimate report to be submitted before December 5th to enable the Cotton Production Improvement Stations to supervise production in different counties and villages at any time (Agriculture Department, 1937d).

Cotton Inspection Stations were set up in the counties with responsibilities for cotton classification, ginning, packing, seed selection, and quality checks on sales cotton while banning illegal trade. Assessment standards for inspectors working at the Cotton Inspection Stations were also formulated: chief inspectors and directors shall be jointly liable for anyone not strictly conducting checks resulting in nonconformant cotton exports; apart from severely punishing the lawbreaker in accordance with the laws, for any bribery and other serious dereliction of duty, the county magistrate or chief county official shall be held responsible for oversight; documents received shall be dealt with in a timely manner without any backlog and delay; staff members of the Cotton Inspection Stations shall not blackmail or bully merchants and farmers by taking advantage of their posts and punishments shall be given to cotton inspectors who failed to meet assessment standards (Agriculture Department, 1936a). In 1937, cotton inspectors of the Cotton Inspection Station in Yicheng County, Erchang Wen and Zhenyin Chen, who received low assessment grades, were suspended from duties for six months by the Improvement Station for Cotton Products of Shanxi Province pursuant to Provision 5 of Article 3 of the Rules for Rewards and Punishments in the Assessment of Cotton Personnel in Shanxi Province (Commercial Bureau, 1937, pp. 17–18).

In addition, temporarily inspectors would be dispatched by the Improvement Station from time to time to deal with special matters. In 1937, as a part of the efforts to promote cotton cultivation technologies, Shanxi provincial government set up trial cotton fields in each counties of Jiyan and offered indemnity guarantee. A technician would be hired for each cotton field to give instructions on a salary of 10 Yuan per month which was approved and paid by the provincial government. In an effort to ensure work efficiency and actual employment and to prevent cheating, inspectors would be assigned by Improvement Station to trial cotton fields in different counties to record workers' names and payroll details to determine the amount of salaries that should be approved and paid (Agriculture Department, 1937a).

IMPROVEMENT AND PROMOTION OF INTRODUCED COTTON VARIETIES

Introduction high-quality cotton varieties and expanded cultivation by Shanxi Province was not simple copying. One of the main purposes of establishing experimental cotton farms is to improve and adapt the introduced varieties to local climate and soil conditions in Shanxi. Much attention was given to improving these introduced varieties. The government placed emphasis on actual results and made careful arrangements in this regard.

The Provincial Government 7th Report on Ten-Year Construction Plan of Shanxi stated that since climate conditions varied from place to place, cottonseeds cultivated in Linfen experimental farm did not necessarily grow in other counties. Therefore, economic cotton cultivation experimental plantations were established in Taigu, Wenshui, Dingxiang and Gaoping counties in ROC Year-10, exclusively for testing high-quality varieties suitable for the different locations. Each farm would lease 100 acres as a cotton planting plot and the farmer was obliged to tour to counties in the vicinity to give instructions (Agriculture Department, 1935b).

Various research projects were launched by the branch experimental plantations in 1935 included: the relationship between climate and cotton crop's pollination period and fertilization; nutrient needed for cotton crop cultivation; causes and prevention of cotton anthracnose; prevention and treatment of aphids; and impact of soil and fertilizer on cotton output. From April to September, a comparative test among U.S. King cotton, U.S. Wood's cotton, Yangqu King cotton, Linfen King cotton, and Wenshui cotton was conducted; seed cultivation tests on King cotton, Wood's cotton, Torres cotton, Yunong cotton, Huilu Dashuo cotton, and Muzi cotton were carried out; and a variety of cultivation tests on King cotton, Trice cotton, and other varieties were performed. Moreover, tests on fertilizer, spacing, steeping, reproduction, boll selection, and plant selection were also conducted (Agriculture Department, 1935c).

In 1935, cotton grown in Dingxiang county of Shanxi Province from introduced cottonseeds was capable of spinning 32 or 42 yarns. When this variety was planted again in 1936, spun yarns reduced to only 22 to 23. Governor Yan Xishan personally called and ordered the Cotton Inspection Stations to study methods for improvement. In March 1937, the government of Shanxi ordered the Improvement Stations for Cotton Products to present causes of the deterioration in Dingxiang cotton quality and come up with remedies.

> Researches and discussions for improvement should be carried out at immediately and it is firmly requested that each cotton farm must actively cultivate and promote fine quality cottonseeds that are drought-resistant, early-ripening, high yielding, and excellent long-fiber pure cottons. Offices in every inspectorate should make earnest efforts in promotion and give practical instructions to achieve practical results as the top priority. (Agriculture Department, 1937b, p. 88, author translation)

Weather conditions in Shanxi are mainly that of the temperate continental climate. Dryness and little rainfall in springtime have an impact on the spring sowing in the cotton fields. Through years of research, the Improvement Station for Cotton Products created a set of effective cotton cultivation methods designed for the climate features in Shanxi—the method of cotton plantation during spring drought consisting of seven methods: namely, the method of digging trenches to catch rainfall, soil-digging method, deep-sowing method, trench-watering method, pit-watering method, ridged row planting method, and transplantation method, which could ensure seed survival under conditions of poor soil moisture. The method was introduced across the province commencing from 1937 and proved to be effective (Agriculture Department, 1937c, pp. 25–26).

Promotion of cotton plantation was mainly centered on two aspects. Upon the completion of cottonseed improvement, the first step was experimental planting; promotion would be the next step after seeing positive results. As a result, after 1936, the government of Shanxi allocated special funds to the counties to set up government-indemnified trial cotton plantations to further popularize the research results from cotton experimental plantations.

Government indemnity measures stipulated specific ways to settle claims regarding trial cotton plantation:

1. Losses incurred by the farmer should be calculated and indemnified based on the prices of regular agricultural products per acre grown in neighboring fields that share the same conditions as the cotton experimental plantations.
2. Due to unfavorable climate, cotton products are sold at prices lower than major agricultural products in the vicinity or crop failure.
3. Those who experienced seedling damage or changing to other crops before germination sell output at lower prices than major agricultural products in the vicinity.
4. One or more types of other crops are replanted due to partial germination, with subsequent combined total incomes from sale of such products lower than other major agricultural products in the vicinity.

5. Cotton produced from cotton fields affected by disasters is sold at a price lower than other major agricultural products in the vicinity. This is not applicable to natural disasters when flooding inundation affects cotton fields and other major agricultural products alike and causes crop failure (Agriculture Department, 1937f, pp. 128–129).

Indemnity scope and compensation standards were strictly defined. This measure aims for experimental cultivation prior to the large-scale promotion of cotton and subsequently developed varieties suitable to the natural environment in Shanxi Province in an effort to observe the output of new products and insect resistance so as to further make improvements that can help reduce crop losses subsequent to promotion and rapidly improving yield. Because plots in the main experimental cotton plantations and branch plantations are limited, the government played a leading role in trial plantation by providing funding while cotton farmers provided land and labor. To prevent cotton farmers from refusing to plant cotton out of concerns regarding output, the plantation stations adopted measures to encourage cotton planting and various means to institutionally stop any attempt to defraud compensation. In 1937, the government of Shanxi Province dispatched special staff to verify compensation for losses suffered by cotton farmers in indemnified experimental plantations in each county.

According to Table 9.2, the government compensated a total of 33 indemnity experimental farms in 1937 alone, involving 447 cotton farmer households in 15 counties and compensation amount of 1,684.4725 yuan, on average paying each household around 3.77 yuan, with highest payment at more than 300 yuan and the lowest at roughly 0.03 yuan. The Compensation Measures stated that this task was important for government reputation and practical results were imperative (Agriculture Departmente, 1937, pp. 105–106).

Secondly, technicians were trained on cotton cultivation. In 1937, the government of Shanxi Province ordered the Provincial Department of Agriculture and the Improvement Stations for Cotton Products to recruit technicians through public examinations and to conduct roughly two-month-long training courses including subjects on seed cultivation and breeding improvement for technicians in farms and cotton laboratories in each county (Agriculture Department, 1937g, pp. 109–111).

Thirdly, cotton sales professionals were trained and the quality of cotton was tightly classified to ensure that cotton merchants were able to understand the classification standards. Counties in Shanxi ordered villages where local cotton farmers had ginning co-ops to assign one to three educated clerks aged 18–30 to receive trainings on cotton classification standards. Inspectors from county cotton experimental stations, construction section

TABLE 9.2 Summary of Farmers' Losses in Indemnified Experimental Plantations Compensated by the Government of Shanxi Province in 1937 (Unit: Yuan)

County	Number of plantations	Households that suffered losses and amount of losses in indemnified experimental farms				
		First	Second	Third	Fourth	Amount
Xing County	3	0	1	10	—	6.098
Qin County	2	5	1	—	—	26.27
Xiaoyi County	2	20	19	—	—	129.4375
Guo County	3	11	17	8	—	152.101
Liao County	1	8	—	—	—	12.133
Taigu County	2	2	3	—	—	11.42
Changzhi County	4	13	4	17	23	198
Huguan County	2	6	10	—	—	36.914
Jiexiu County	2	54	9	—	—	427.42
Jiaocheng County	2	6	16	—	—	75.54
Pingyao County	1	5	—	—	—	20.1
Pingshun County	1	15	—	—	—	25.199
Yu County	3	0	0	58	—	96.43
Shouyang County	2	21	28	—	—	164.61
Lishi County	3	23	19	15	—	302.8
Total	33	189	127	108	23	1684.4725

Source: Agriculture Department, 1937b, p. 88

chiefs from county governments, and technicians took on the role of trainers while county chiefs were responsible for supervision and guidance. Training courses included abstracts of cotton administration measures, cotton classification and test internship, overview of cotton crop management, practical guides on cotton classification and water content and impurity assessment, and elementary introduction to the cotton industry and crop management. Training fees were covered with the fees collected by the ginning co-ops (Agriculture Department, 1937h, pp. 51–52).

TECHNICAL IMPROVEMENT AND COTTON YIELD AND QUALITY INCREASE

In late Qing Dynasty and early years of Republic of China, affected by local customs, "most of people use fur and leather as clothing and bedding" in Shanxi. Given that local people had little demand for cotton, cotton output was mostly "shipped and sold to other provinces, weighted in about two hundred thousand quintals and valued about 7 to 8 million Yuan. In addition to sales to Tianjin, the cotton was also sold to Shanghai and Wuhan" (Industrial International Trade Bureau, 1937, p. 474, author translation). As Shanxi is an inland province, cotton production is not a traditionally favored activity. The essential reason for the ROC Government of Shanxi to promote cultivation of cotton is that for China's exports, "cotton yarn, cotton cloth, woolen and carpet produced by the textile industry are among the much well-known products sold in domestic and international markets" (Industrial International Trade Bureau, 1937, p. 801, author translation).

In 1917, a cotton experimental plantation was established. In the following year, the planting area of cotton in Linfen, Shanxi increased to more than 200,000 acres and the yield of lint reached more than 300,000 quintals. The planting area increased to 486,000 acres and 835,700 acres in 1919 and 1920, respectively. In 1921, another cotton experimental plantation was set up in Jiyan District and by the end of 1920, the region's cotton production increased to over 1,449,180 kg and 17,357,175 kg in Hedong District. The province's total output exceeded 18,806,360 kg.

Shanxi provincial government vigorously promoted and improved cotton cultivation technology. In 1927, cotton production in Linfen and 20 counties in southern Shanxi stood at 3,077,752.5 kg, valued at 1,470,467 silver dollars (see Figure 9.1). With Ruicheng as the center and a 100 km radius around, cotton output in Yongji County, Yuxiang County, and Pinglu County

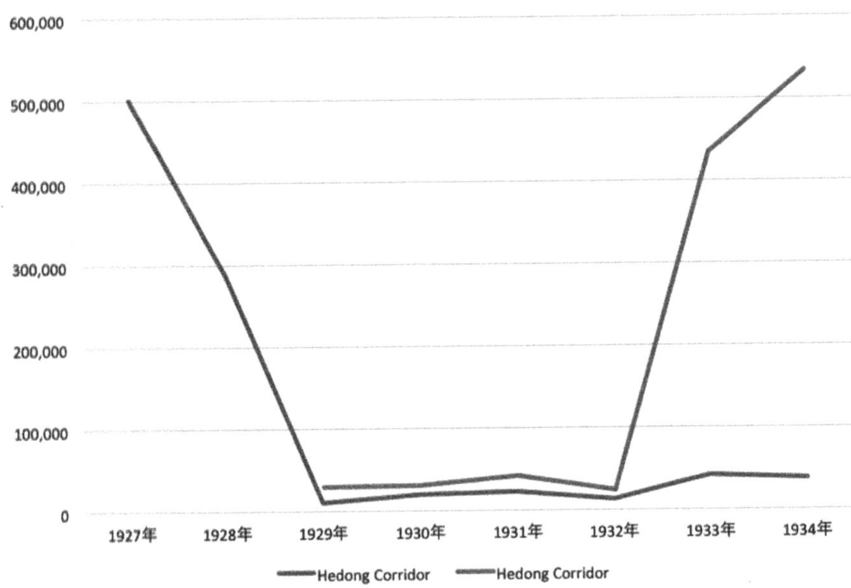

Figure 9.1 Cotton field and cotton yield growth in the counties of Shanxi Province.

came in at 360,650 kg. Within a 200 km radius, cotton output in Quwo County, Yicheng County, Fencheng County, Xiangling County, Ronghe County, Wanquan County, Xinjiang County, Jishan County, and Hejin County stood at 2,381,566.5 kg, accounting for about 62.6% of the total output of Shanxi Province (Shanxi Provincial Government Statistic Bureau, 1930).

In 1935, the cotton-producing area in Shanxi expanded, giving the province the 8th ranking, following Hebei, Jiangsu, Hubei, Shandong, Henan, Shaanxi, and Fenshui Basin in Sichuan (Xiao & Zhang, 1937). As cotton production gradually increased, by 1927, Shanxi shipped 5,989,941 kg of cotton worth 4,640,918 yuan to other provinces through the Shijiazhuang-Taiyuan Railway. At the same time, cotton output in Fenglingdu was 124,591.5 kg, valued at 79,844 yuan (Shanxi Provincial Government Statistic Bureau, 1930). Influenced by the introduction of U.S. varieties, production of local varieties in Hedongdao started to fall steeply starting in 1927, and by 1934, it fell to less than 7% of the original output. On the contrary, 1934 output of overseas varieties in Hedongdao increased by approximately 18 times the 1930 level. Production of local varieties in Yanmendao during a five-year period from 1927 to 1931 was less than a third of the output in Hedongdao for one year in 1929, while this year saw the lowest output in Hedongdao. Yanmendao began to plant overseas varieties starting in 1931, and production increased slightly, but the rate of increase was far behind the outputs of the other two areas. Jiningdao is located in central part of Shanxi Province and it is not a center of cotton production; therefore, the cotton production here was not affected by the American varieties while output of local varieties had not increased even in the face of soaring demand. Cotton production was not blindly expanded in this region (He, 2001).

The soil and climate conditions of Linfen, Puzhou and Xiezhou, Shanxi Province are suitable for planting cotton. Since the late Qing Dynasty, these regions have been cotton-producing centers in Shanxi. Due to the lack of knowledge of the cotton market, cotton did not arouse the attention of growers and the provincial government. In 1917, Shanxi provincial government started to become aware of the importance of cotton to the national economy and after establishing a cotton experimental plantation in Pingyang, the cotton quality in Linfen, Shanxi greatly improved. In 1918, cotton shipped through Shijiazhuang-Taiyuan Railway to Japan was identified by Japanese importers as the "most superior fine fibers of the country" (Agriculture Department, 1981, pp. 174–175, author translation).

After vigorous promotion of cotton, especially after Shanxi Province formulated its Ten-Year Construction Plan and as output rose, cotton quality was safeguarded by two government measures of turning attention to setting the cotton grading standards and presenting awards for high-quality varieties.

In 1936, in order to improve the quality of export cotton, to enhance to reputation of export cotton, to improve quality of cottonseeds and cotton products, and to prevent shoddy products and adulteration with water and impurities, Measures for Administration of the Cotton Industry was issued by the Shanxi provincial government setting the grading and calibration standards (Table 9.3) for cotton quality and also stipulating relevant provisions on cotton ginning, packaging stamps, and so on.

A cotton industry inspection bureau was established to supervise cotton ginning and packing and to conduct cotton inspection, cottonseed tracking, and other matters. Inspectors and officers were dispatched to major ports of trade for cotton to check the quality of imported and exported cotton products. Relevant provisions were stipulated on violation of Cotton Industry Administration Measures. Minor offenders would be subject to fines ranging from 1 to 1,000 yuan while serious offenders would be handed to the local judiciary authorities and dealt with in accordance to the law (Agriculture Department, 1936c).

In addition, to encourage farmers to improve the quality of cotton, Shanxi provincial government also stipulated that Hedong and Jiyan Districts should hold cotton product exhibitions after the autumn harvest each year. The cotton experimental plantation together with local county government should hold cotton quality assessment sessions to review the

TABLE 9.3 Cotton Grading Standard Specified by Shanxi Province in 1936

Variety	Grading Standard	Grading	Standard
U.S. cotton	Coloration	Superior	Bright and gleaming, unblemished and pure white, shiny
		High	Creamy white, unblemished
		Middle	Yellow white, yellow inside. Cotton fibers does not exceed 50%
		Low	White, yellow inside. Cotton fibers does not exceed 50%
		Poor	Grayish white, yellow inside. Cotton fibers does not exceed 50%
	Fiber length	High	Longer than one inch
		Middle	Shorter than one inch, ¾ inch and longer
		Low	Shorter than ¾ inch
	Fiber neatness	High	90%
		Middle	85%–89.9%
		Low	75%–84.9%
Chinese cotton		High	Pure white, unblemished quality, fiber length over ¾ inch
		Middle	Yellow, unblemished quality, fiber length over ⅗ inch
		Low	Red, slightly contaminated, fiber length over ½ inch

Source: Agriculture Department, 1936b

progress status on production policies and select farmers' planting high-quality cotton varieties to be recommended to the provincial government for an award (Historical Annals Research Institute of Shanxi Province, 2001, pp. 121–122).

To avoid possible malpractice in the assessment session, Shanxi provincial government set strict requirements for the organizing body and the assessment method of the cotton industry assessment session. First, the address and date of the assessment should be submitted by the cotton experimental plantation during preparation to the Industrial Department for provisional approval; the executive agency of the assessment session should be composed of a chairman, a president of supervisory council, plus several numbers. The director of the industrial department should act as chairman; the head of the cotton experimental plantation should act as president of the supervisory council, and members of supervisory council should be temporarily appointed by the chairman and president of council; an experts penal should consist of one assessment chairman and several assessment members to be provisionally designated by the chairman and president of council (Agriculture Department, 1934). Specific requirements were stipulated on appraisal of cotton quality including luster, coloration, fiber strength, callus, cottonseed weight, and fiber length (See Table 9.4).

The total score of products should be the sum of scores by visual and instrument inspections divided by 2. Given that cotton industry promotion in northern Shanxi started later than Linfen and due to poor natural conditions in this region, the requirement "when making assessment, products from Jinyan District should be given 10% score higher than products from Hedong District" (Agriculture Department, 1934, author translation). Famers being given scores of 60 or higher would receive an award: Producers scoring 60–80 would be considered as middle to lower rating, 80–90 as medium rating, and 90–100 as excellent. Medals and certificates would be presented based on ratings (Agriculture Department, 1934).

CONCLUSION AND FUTURE STUDY

High-yield cotton has created economic benefits for Shanxi Province by providing raw materials for other major cotton-processing provinces. Such profits have been considerable after deducting freight, which also contributed to development of the cotton machinery manufacturing industry in Shanxi Province. Take cotton packing, for example. Initially, the sites for cotton ginning and packaging were selected by each county, and the cotton traders autonomously grouped together for centralized packing; the packed bales would then be transported to larger towns for inspection. Each cotton trader was required to have a ginning machine with the

TABLE 9.4 Cotton Quality Appraisal Standard in Cotton Industry Assessment Session

	Item	Scoring	Total Score
Visual inspection	Sheen	Presence of fiber sheen as compliant to Standard. Scores are based on brightness.	20
	Coloration	Pure white as compliant to Standard. Scores are based on coloration.	20
	Tensile strength	Elasticity as measure of tensile strength. Scores are based on degree of elasticity.	20
	Callus	Even fiber thickness and no white spots as compliant to Standard. Scores are based on the number of white spots.	
If total score of above four items is less than 60, instrument inspection shall not be conducted			
Instrument inspection	Cottonseed weight	10g as compliant to Standard. Scores are based on the weight	20
	By percentage	35% as compliant to Standard. Variation of percentage point equals variation in score point.	20
	By average weight	10g as compliant to Standard. Variation of percentage point equals variation in score point.	20
	By net seed weight	20g as compliant to Standard. Variation of percentage point equals variation in score point.	10
	Fiber length	3cm as compliant to Standard. Each unit variation in length equals variation in one score point.	20

Source: Agriculture Department, 1934

capacity of 4,500 catties or more of mill cotton per day and a packaging machine with the capacity of 100 bales or more per day (each bale weighing approximately 150 catties). Producers who could not afford the machines were allowed to rent or share the machines; otherwise, their cotton merchant business license would be revoked (Agriculture Department, 1934). In 1936, in order to unify the standards for bulk cotton buying, simplify procedures for bulk cotton buying, and increase cotton export capacity, Shanxi's provincial government began to select appropriate sites within the province to build high-capacity cotton packing factories. Additionally, Bin Kee & Co., under the Board of Public Enterprises of Shanxi, contacted five British and German firms including Jardine, Conn Matheson, Siemssen & Co., Carlowitz & Co., and Symon's to negotiate the purchase of a cotton baler and related technologies. This led to further development of modern industry in Shanxi Province.

REFERENCES

Agriculture Department. (1934). Shanxi cotton industry assessment pamphlets. *Government Gazette of Shanxi Province Bulletin, 30,* 32–33.

Agriculture Department. (1935a). Cotton industry special project in the ten-year construction plan of Shanxi Province. *Construction in Shanxi Province Bulletin, 10,* 1–5.

Agriculture Department. (1935b). Provincial government 7th report on ten-year construction plan of Shanxi (progress in cotton promotion). *Construction in Shanxi Province Bulletin, 6,* 4–8.

Agriculture Department. (1935c). ROC Year-24 implementation measures for branch experimental cotton plantations in Shanxi Province. *Government Gazette of Shanxi Province Bulletin, 38,* 36–39.

Agriculture Department. (1936a). Service rules for personnel of county cotton inspection stations of Shanxi Province. *Government Gazette of Shanxi Province Bulletin, 52,* 38–39.

Agriculture Department. (1936b). Shanxi Province cotton industry administration measures. *Government Gazette of Shanxi Province Bulletin, 52,* 22–23.

Agriculture Department. (1936c). Shanxi Province cotton industry administration measures. *Government Gazette of Shanxi Province Bulletin, 52,* 34–35.

Agriculture Department. (1937a). Government order of Shanxi Province (Jian Nong Zi No. 132). *Government Gazette of Shanxi Province Bulletin,* 62–63.

Agriculture Department. (1937b). Government directives of Shanxi Province (Jian Nong Zi No. 645). *Government Gazette of Shanxi Province Bulletin, 13,* 88.

Agriculture Department. (1937c). Method of Cotton plantation during spring drought. *Government Gazette of Shanxi Province Bulletin, 18,* 25–26.

Agriculture Department. (1937d). Notes for cotton production improvement station village cotton production surveyor. *Government Gazette of Shanxi Province Bulletin, 15,* 27–28.

Agriculture Department. (1937e). Orders of the government of Shanxi Province (Jian Nong Zi No. 13). *Government Gazette of Shanxi Province Bulletin, 3,* 105–106.

Agriculture Department. (1937f). Orders of the government of Shanxi Province (Jian Nong Zi No. 253). *Government Gazette of Shanxi Province Bulletin, 27,* 128–129.

Agriculture Department. (1937g). Orders of the government of Shanxi Province (Jian Nong Zi No. 64). *Government Gazette of Shanxi Province Bulletin, 7,* 109–111.

Agriculture Department. (1937h). Training manual for agricultural and commodity cotton personnel in counties of Shanxi Province. *Government Gazette of Shanxi Province Bulletin, 16,* 51–52.

Agriculture Department. (1981). Linfen cotton experimental plantation. *Industry News, 1,* 174–175.

Commercial Bureau. (1937). Government order of Shanxi Province (Jian Shang Zi No. 105). *Government Gazette of Shanxi Province Bulletin, 11,* 17–18.

Editor of Industrial News Highlights. (1918). Experimental cotton farms in Linfen. *The Industrial Society Magazine, 1,* 174–175.

Editor of Jinyang Daily. (1936). Shanxi Province over the last 30 years. *Jinyang Daily, 71.*

Feuerwerker, A. (1970). Handicraft and manufactured cotton textiles in China, 1871–1910. *The Journal of Economic History, 30*(2), 338–378.

He, S. (2001). Boom and bust of China's cotton textile industry during 1914–1925 by reviewing shift in the supply and demand curves. *Academic Quarterly Magazine of Shanghai Academy of Social Sciences, 4,* 167–175.

Historical Annals Research Institute of Shanxi Province. (2001). Appendix. In *General Annals of Shanxi Province* (Vol. 50, pp. 121–122). Beijing: Zhonghua Book Company.

Industrial International Trade Bureau [China].(1937). China Industrial Annals. Part IV. In *China Industrial Annals.* Industrial International Trade Bureau Press.

Li, L. (2010). The Development of Railway and Modern Cotton Industry of Shanxi: 1907–1937. *Journal of Shanxi Agricultural University (Social Sciences Edition), 3,* 342–345.

Lillian M. L. (1981). *China's silk trade: Traditional industry in the modern world 1842–1937.* Cambridge, MA: Harvard University Press.

Lin, G. (2002). A new exploration into China's handmade cotton and textile industry during 1927–1937. *Chinese Economic History Research, 2,* 17–33.

NO. 3500 department [Japan]. (1942). *The overview of main workshop in Shanxi, the report by Japanese 3500 department in 1942.*

Shanxi Provincial Government Statistic Bureau. (1930). *Shanxi ninth economic statistics.* Shanxi: Shanxi Provincial Government Press.

Statistical Bureau. (1935). Provincial government 7th report on ten-year construction plan of Shanxi (progress in cotton promotion). *Construction in Shanxi Province Bulletin, 6,* 4–8.

Walker, K. L. (1986). *Merchants, peasants, and industry: The political economy of cotton textiles, Nantong County, 1895–1935.* University of Michigan.

Wang, Y. (1984). Research on Shanghai's cotton planting history. *Agricultural History of China, 1,* 38–43.

Xiao, F., & Zhang, L. (1937). Chinese modern cotton industry. *Zhejiang Cotton, 5,* 195–206.

Xu, T. (1998). *The development of Shandong's commodity economy in Ming and Qing Dynasties.* Beijing: China Social Sciences Press.

Xu, Y. (2012). Implementation and performance of cotton governance during 1917–1937 in Shanxi Province. *Journal of North University of China (Social Sciences Edition), 2,* 1673–1646.

Yan, Z. (1943). *Development of China's cotton industry.* Beijing: Commercial Press.

Yu, X. (2010). Study of foreign trade of cotton in Yangtze River delta during. *Researches in Chinese Economic History, 3,* 65–72.

Zhang, L. (1991). Overview of modern cotton industry in Jiangsu province. *The Journal of Chinese Social and Economic History, 3,* 80–88.

Zhuang, W. (1996). Introduction of new varieties of crops by Shandong in modern times and its impact. *Modern History Research, 2,* 189.

CHAPTER 10

MOVING ON LINE

Public and Private University Motivations and the Implications on Public Policy

Michael T. Miller
University of Arkansas

Mei-Yan Lu
San Jose State University

ABSTRACT

This chapter argues that higher education institutions and leaders often do not see themselves as part of a larger system, and with an attention to serving students on their campuses, they often do not need to have this perspective. However, when seen from the larger perspective of serving a state or province, institutions have to look at how they reach out to students, and as they see the benefits and challenges of moving classes into an online format, they see that they are not alone. The policy implications are discussed in details in terms of how institutions have to talk to each other, to assess what others are doing, how they are providing education, and what resources can be maximized across institutions and even public sectors to best serve their public interests.

Technological Innovation Networks, pages 211–229

> Online learning has matured greatly in the past decade, but is still in its infancy in terms of integration to the larger perspective of higher education as a public service, and the next two or three decades could rewrite how colleges and universities provide a postsecondary education.

Colleges and universities around the globe are reconsidering how they are structured, what they are designed to do, and how they perform the tasks they are assigned to do. Whether public or privately funded, institutions are exploring efficiency and outcomes as never before, and even federal governments, including the Obama administration in the United States, are forcing these educational institutions into tightly defined areas of responsibility. The Obama policy of rating higher education institutions, for example, includes specific elements of job placement upon graduation, a clear sign that these leaders expect colleges and universities to be a major component in labor preparation.

College leaders have struggled to accept such a singular perception of their work, yet they have largely discarded philosophical conversations about the purpose and meaning of higher education (Bowen, Schwartz, & Camp, 2014). In the United States, state legislatures and policymakers have further limited higher education's ability for self-manifestation by creating performance funding systems that mandate the measures by which institutions must perform. Such measures, whether job placement rates, graduation rates, or average student debt, can be seen as a tool for state policymakers to keep institutions in line with their public purpose (Burke, 1998). The difficulty of such funding models, however, is that they define a portion of the institutions' activities without a concurrent amount of funding accompanying the mandates, nor do they concurrently reflect governance structures that support such activities. For example, a typical U.S. higher education institution receives about half of its revenue from public funds, yet 100% of the institution's governing control comes from the public sphere (Dougherty, Natow, & Vega, 2012).

Philosophical discussions of the meaning and purpose of higher education have moved consistently toward the economic benefit to regions, people, and economies of postsecondary education (Bowen et al., 2014). The economic benefits are both direct, meaning individuals are trained for jobs that pay them more money, and indirect, meaning that individuals learn about life choices that result in both a higher quality of life and fewer unhealthy behaviors and subsequent health-related illnesses. These discussions are not limited to the U.S.; Holmwood (2015), for example, noted in the United Kingdom that 70% of the public believes that there are more advantages to a university education than financial gain. As these philosophical discussions are truncated in the world of legislative and public discourse, they are also simplified and directed at measurable outcomes, resulting in the reduced emphasis on personal student development.

Student development as a purpose of higher education has played an important, indirect role in the history of higher education (Karabell, 1998). Although the formality of assessing and stressing student development has emerged during the past 50 years, it has come to represent the personal journey a young adult experiences while enrolled in a learning environment. A wide variety of psychosocial theories work to explain the confrontation with crisis and the emotional and psychological result of responding to these crises resulting in an individual who has a better sense of self-purpose and direction. In the realm of student development, activities such as social gatherings, sports teams, and common living and dining areas all have purpose and value; they create community and allow individuals to test social interaction and their own beliefs (Astin, 1984). Historically, then, the individual who has had such experiences is best prepared to emerge from postsecondary studies to take on professional responsibilities in the workforce and have that preparation represented in higher working wages. At some point in the not-too-distant past, this developmental function of the academy seems to have been discarded for the ends-oriented belief that college equates to job training and higher working wages.

This societal shift in thinking about college as a preparatory experience for the workforce and for personal financial gain (Hersh & Merrow, 2006) has shifted cultural thinking about the appropriateness of short-term financial loss for the potential long-term advantage of higher working wages. One result of such thinking is that institutions can profit from a student willingness to go into debt and pay more for a degree. One segment of the higher education industry to seize on this willingness has been proprietary institutions who can make a profit from student enrollment, and these institutions average significantly more in what they charge for tuition than their public, nonprofit counterparts. And these institutions' desire to earn profits also results in their willingness to take less well-prepared students and their exploration of alternative academic program delivery.

The growth in proprietary student enrollment, from under 800,000 students in the U.S. alone in 2001 to over 2.5 million in 2010, suggests a more competitive enrollment management environment where public and private institutions actively compete for many of the same students. This competition includes not only lifestyle issues of the college campus, but also convenience factors related to time-to-degree, employability of graduates, and convenience of classes offered. Repeatedly, the movement to offer online programs has come through a desire by institutions to follow student tuition dollars (Miller & Morris, 2008).

Academic program migration to online platforms for financial gain has legitimacy, especially for niche institutions that rely on tuition dollars as their primary source of revenue (Miller & Morris, 2008). Online programming, however, also has relevance in the discussion of how public institutions serve their constituents, making higher education available to larger segments of the population who are bound by geography. Additionally, online learning has been shown to be highly effective and efficient in reaching students and providing them meaningful learning (Topper, 2007).

Research about the effectiveness of online learning has been inconsistent, with some findings suggesting that online courses that embed interactivity are highly effective, while other research has shown that it is a poor substitute for face-to-face instruction (Chapman & Henderson, 2010; Dykman & Davis, 2008). The general consensus, however, among academic leaders is that using online instruction can be an important element in institutional offerings that expands access to postsecondary education while simultaneously providing additional student tuition dollars at a critical time for higher education (Mintz, 2014; Selingo, 2013).

Based on these trends of growing to rely on online student tuition dollars, the need for reaching geographically diverse students, and the debates about the effectiveness of online learning, the current chapter was designed to explore both what is happening within the growth of online learning and what kinds of policy implications arise from this new attention to educational delivery. Although the discussion of online learning policy issues is primarily situated within the United States context, the implications are similarly felt and realized throughout many nations and in many governments.

MOVING INTO THE ONLINE WORLD

The rationale for colleges and universities to offer programs in online formats has typically fallen into three categories: to preserve or strengthen program enrollment, to generate revenue, or to provide access to underserved populations or regions. The first two of these reasons are situated in the paradigm that enrollment results in financial gain, and with fewer state dollars feeding public institutions and rising costs at private colleges, enrollment dollars prove to be precious (Miller & Morris, 2008). The idea of serving underserved populations has also been seen as a potentially financially lucrative proposition for institutions; however, the motivation is really about helping those in distressed or highly inaccessible locations have access to education.

Whatever the motivation, online learning is growing rapidly. In 2012, about 5.4 million college students, approximately one in four, took at least one online college course (see Table 10.1). Student motivation for taking online courses includes the convenience of learning whenever and wherever they might choose, availability of specialized programs otherwise not

TABLE 10.1 Online Student Enrollment in the U.S. by Institution Type

	Exclusively online	Some online
Undergraduate Enrollment	2,002,815	2,582,475
Public		
4-year	574,709	1,223,442
2-year	674,134	1,182,801
Private nonprofit		
4-year	466,730	257,599
2-year	798	2,198
Private for-profit		
4-year	901,601	121,684
2-year	21,711	19,749

Note: less-than-2-year institutions were included in grand totals, but not in break-out of type of institution.
Source: NCES, 2014.

available, the ability to transfer into other institutions, cost, and even learning style preference.

College senior academic officers further reinforce online education as important to their financial well-being and their place in the postsecondary market, and nearly two-thirds (70.8%) report that offering online education programs is critical to their long-term success and strategy (Allen & Seaman, 2015). Despite this, however, faculty have not accepted the online movement with open arms, as only 28% of senior academic affairs officers have indicated that their faculty accept online learning as being of value and legitimate. Almost as many senior academic officers (25.9%), however, have indicated that they believe online learning is "inferior" or "somewhat inferior" to face-to-face instruction (Allen & Seaman, 2015).

Despite the growth and popularity of online courses in recent years, students who enroll in them do not complete these courses particularly successfully and at the same rate as students who enroll in face-to-face courses (Allen & Seaman, 2010, 2015; Layne, Boston, & Ice, 2013). This can lead to drop-out rates for students enrolled in online courses six to seven times higher than students enrolled in face-to-face courses (Patterson & McFadden, 2009). This higher drop-out rate could be attributed to the fact that students need more discipline and support to succeed in an online course than in a face-to-face course. Higher education academic leaders' reporting that students need more discipline to succeed in online courses has remained consistent over time (Allen & Seaman, 2015).

A different concern for the growth and future of online programs is a concern among senior academic leaders is the potential of lack of acceptance of online education by potential employers. Nearly one-half of all leaders among the for-profit sector rank this as a critical barrier (47.2%) while 37.4% for leaders at private, not-for-profit institutions and 32.8% of the leaders at public higher education see it as a barrier (Allen & Seaman, 2015).

To maximize cost-effectiveness and potentially work with faculty in a different way, some institutions have made use of massive open online courses (MOOCs). A MOOC is an open online course that any person can take, frequently comprised of readings and recorded lectures that have no limit on attendance, are free or are of very low cost, and are generally, but not exclusively, offered by proprietary providers. Despite their ability to cheaply deliver mass education for introductory level subjects, and a growth of nearly 5% of institutions that use MOOCs (up to 8%), relatively few colleges and universities have committed to using them in the future (Allen & Seaman, 2015).

IMPLICATIONS FOR POLICY

State Policy on Postsecondary Education

In the United States, each individual state maintains a right to control the offering of higher education within state boundaries (Sibley, 1998). This means that some form of governmental agency either coordinates or governs the postsecondary education industry, and that these bodies are either appointed, elected, or a combination of both, with the power to ultimately decide which institutions operate in their states. About half of all U.S. states make use of a coordinating board with regulatory authority over higher education (Heller, 2004).

The concept of operating within a state is based on a 20th-century mentality that a physical campus is necessary, with an instructor present in a set location delivering some form of lecture or teaching a class. This conception was rarely challenged in the era of correspondence courses or even in reference to instructional television, but the popularity of online learning has created a significant discourse at the state level about how to allow and approve an online course that originates at a distant location.

In addition to state authority granting for institutional operations, individual institutions and state systems have had to begin dealing with how they regulate offerings within their own boundaries. Many comprehensive university systems and community colleges, for example, have defined service areas based on the notion of a physical presence. Much like local school districts, these colleges are charged with providing educational programming to the district, but with online course offerings available, district boundaries become incapable of restricting who can access individual courses or programs of study.

Individual states have strong motivations to regulate higher education within their borders, as student spending on tuition is significant and states have a de facto responsibility to assure that credentials issued within their boundaries have an appropriate level of merit and authenticity. State bureaucracies vary greatly in their interpretation of how involved to be in regulating the higher education industry, with some states requiring substantial documentation of faculty credentials and personal meetings with institutional officials, ranging to brief applications that are approved without question as long as an appropriate filing fee is paid (U.S. Department of Education, 2012).

Philosophical thinking about which higher education institutions can operate within state borders can be driven by political thinking, with policy resulting from negotiations between competing interests. In the current U.S. political environment, a competitive market-economy perspective has allowed institutions to open in record numbers throughout the country. The conception that drives this thinking is that competition is good, and competition between educational providers, both online and face-to-face, will result in the survival of high-quality institutions and the closure of those that lack the credibility to continue. Such thinking often results in less government regulation in some state legislatures and cooperatives of states in other settings.

The most popular consortium of state level approval for online education offering is the National Council for State Authorization Reciprocity Agreements (NC-SARA). This compact of states was driven largely by the federal government's call for online course accountability and recognition and approval for operation within a state in order for students to receive federal student financial aid money. This effort to streamline state-level recognition among educational providers is based on a proposed single filing and determination of quality, followed by an acceptance by state departments of higher education that the provider is of appropriate quality and rigor and is automatically granted permission to offer online courses to citizens of a given state (NC-SARA, 2015).

There are problems with NC-SARA, however, as individual states see participation in such a reciprocal agreement as potentially a loss in revenue, an increased competition with their own state-sponsored programs, and differentiation in standards due to regional differences, institutional quality, and instructor quality. The default conversation has gravitated to Quality Matters, a private organization that purports to have stringent guidelines for what should be embedded in online classes and how online classes should be structured. Quality Matters relies on a scoring rubric for individual classes along with peer review, both of which have been criticized for violating the academic freedom of faculty members.

Additional challenges to state-level policy for online learning have to do with the regulation of content delivery—namely, the identification of what constitutes credit hour production in an online course. Traditionally, course content delivery was regulated based on the number of seat-hours, that is, the actual length of time that a class would meet. Without parameters for how long a student "sits" in class, states have difficulty regulating the seat-time equivalency (Ginn & Hammond, 2012).

From a state policy perspective, however, online education is viewed as a commodity to be regulated and bargained with among peer state authorities, much the way online commerce has been structured. And although the general idea of competition among higher education providers is seen as a good practice, the potential decrease in revenue is seen as a motivator for states to build walls preventing other providers from entering their state boundaries.

Workforce Education

Workforce development education is a process of preparing individuals for the world of employment. This type of education can take place in secondary schools, postsecondary institutes, or higher education; it can be offered by state agencies or private providers, and it can include specific programs of occupational preparation as well as adult basic education. As programs are moved into online formats, there are multiple levels of questions that arise for state policy (Olesen-Tracey, 2010).

In the very recent past, the U.S. moved its national high-school equivalency education program, the General Education Diploma (GED), from a nonprofit based provider to a private-sector, proprietary company, Pearson Publishing. Pearson immediately set out to bring the GED test to current secondary school level national standards, complying with the U.S. Department of Education's Common Core requirement. The GED test was then keyed to these skills, with passing scores aligned to the knowledge required of a 12th grade student, in addition to placing the test in a computerized format and requiring test proctors to be certified with certain credentials (Cody, 2014). These changes were radically different from the paper-and-pencil test that was not calibrated to any particular state or national standard. So as the test becomes more consistent with thinking about what is required of a high-school level education, states have had to address how they can most effectively reach those without a secondary credential (high school diploma). In virtually every state, state departments of education have made it a policy to offer secondary school training programs to drop-outs through virtual formats, offering these programs primarily but not exclusively in online formats. The resulting policy shift is from funding localities to provide this training and test proctoring to state-level agencies providing mass online learning and testing. This creates massive cost savings and standardizes curriculum, but it also fails to recognize differences in geography and learner needs.

Similar state-level policy discussions are focused on these virtual academies and training programs that offer certificates, diplomas, and similar credentials based on the completion of an online program. Some route a variety of job training through state-level providers, such as a department of labor or human services, and others have created virtual high-schools through their departments of education or similar proxy. These online providers are then capable of providing concurrent occupational training for secondary school students while they are still enrolled in their high school classes. Model programs often accompany online learning programs with locally sponsored apprenticeship programs, allowing students to earn a high school degree as well as a license of practice for the specific occupation.

Similar online programming efforts have been coordinated at state-department levels where the emphasis is on career development. Some states, for example, provide free career exploration online courses and programs to high school students, offered to individual high schools or to high school consortia. These programs have also included language and cultural instruction for regions where foreign employers have located, allowing students to develop a cultural and language familiarity that can aid them in transitioning to work in these facilities.

Perhaps the biggest effort to use online education for workforce development has been in the general area of critical thinking, practically demonstrated with the completion of a high school (secondary) diploma. State departments of education have in particular worked to use online education tools to provide high school opportunities to a wide range of potential students (traditionally aged students as well as older adults), in some cases charging a specific high school district to provide such services; in some cases chartering a state-wide virtual high school; and in some cases partnering with universities to provide online education. Additionally, state departments such as education, labor, human services, and economic development have teamed up to offer high school level education through online formats, delivering programs at discounted user fees or free of charge, offering tutoring, mentoring, and academic skill support through business and industry. These partnerships encourage workplace learning of basic skills without interrupting the worker/student's time on the job (Taylor, 1984).

The return on investment for state policy to encourage online learning is that the more educated workforce, one that can be documented through the number of diplomas or degrees awarded, can attract more and higher-quality businesses and jobs to a given state. In many southern U.S. states, this might mean that car manufacturing or any sort of industrial manufacturing is more likely to locate there because there is an educated workforce ready to be employed, and online platforms allow this to happen in timely, efficient, and economically pragmatic ways.

Labor and Training

The majority of labor and training related programs coordinated or offered through higher education institutions are those that are funded, sponsored, or requested by private business and industry. Often, these programs take the form of short-term contract training courses, where a given local business contracts with the local college (community college or university) for a certain type of training. The college can then make use of its own resources, including facilities and faculty, or it can subcontract with private individuals to provide the training. Increasingly, as industry specializes, seeks compliance with international standards, and draws upon the global framework of business, colleges find their instructors and faculty unable to offer such specialized programs. The response has increasingly been to utilize online formats to bring experts to campus to provide the training. Additionally, colleges are capable of using their intranet capability of networking multiple business locations with off-site, global instructors to provide such training. Programs such as these are particularly popular in community colleges and short-cycle postsecondary institutions that rely on such training programs as financial buttresses to their traditional tuition based programs.

Other areas of training that have moved into online formats are job retraining programs, skill enhancement programs, lifestyle and career education programs, career reentry, and even second-career, return-to-learn programs. This wide variety of programming continues to be offered in live, face-to-face settings; however, higher education leaders have found greater capacity to offer programs and financial efficiency in using technology to offer programs such as these. In the U.S., at the state level, many of these programs are funded in whole or in part through different agencies, such as departments of labor, education, or human services, and they have a particular interest in programs that support employment. These agencies find the higher education infrastructure a favorable network for distributing such learning programs, and instruction that is captured in a media format that can be redistributed can be financially and technically efficient.

From a policy perspective, online education programs targeted at labor and job training programs can be incentives to keep business and industry in specific locations, namely within state borders, and can be an incentive to help a business or industry relocate to a state. State-level policy can also be used to structure programs such as these for different populations at different stages in their lives, such as retraining for individuals who have lost their jobs due to factory or industry closure, or coupling programs such as these with other social service programs, such as basic literacy education.

The extent that policymakers engage in online training programs as a matter of policy is related to perceptions about governmental support for business development. Conceptually, policymakers are designing programs that improve the lives of individuals, but they both directly and indirectly support the state's business climate. Policy directives that are pro-business tend to focus more on customized training delivered to industry as a benefit of location, while socially oriented policymakers drive programs that focus on personal skill enhancement and career orientation. Higher education leaders have historically accepted either directive, as they benefit from the financial delivery of programs, although they have found that chasing such funding can be damaging to the pursuit of their institution's mission (National Health Service Corps, 1992).

Online program delivery in these types of programs has grown to be of particularly high quality, as external funders generally provide the money to create such classes and curriculum. Attention is often directed at content-specific skills, and the online format allows institutions to produce a class or educational program in a professional manner, using content experts and examples from the global community and involving real-time speakers and content experts in the application of given concepts or ideas. The ability to then repackage some of this training content can allow the aggressive higher education institution or education provider the opportunity to expand program offerings and grow revenue. The advantage can be both to the delivering institution as well as to the local company, giving it a global exposure that it would not otherwise be capable of purchasing.

This approach to business and industry training is far from new to higher education; however, the integration of technology is only now beginning to grow at a pace consistent with its ability. As institutional leaders frame their mission that includes community service, labor and job training programs have become central to their growth. Of increased importance, then, is the communication with state agency personnel to assure an integrated approach to job training that includes multiple providers, including the higher education institution, and highlights the efficiency available through online programs.

Disability

A significant movement in the U.S. and around the world for the past 30 years has been to focus on how to best accommodate individuals with different types of disabilities, also called "exceptionalities." These range from vision impairment to mobility issues, and the attention to their accommodation has resulted in a growing body of policy-related decisions about nondiscrimination (Marshak, van Wieren, Raeke, Swiss, & Dugan, 2010). The U.S. in particular passed a federal law 25 years ago, the Americans with Disabilities Act, which outlined the specific accommodations public agencies must make (Chatterjee & Mitra, 1998). These range from specifically identifying the essential functions of a job and interviewing and selecting candidates based on their abilities to fulfill those functions to requiring physical spaces that allow disabled individuals to navigate these settings. In education, this means that doorways must allow wheelchairs to fit through them, classrooms have brail numbering, sidewalks have graded cut-outs for access, and buildings have accessible mechanisms to carry individuals between floors. All of this gets brought into the discussion of online education in a meaningful way, as institutions can use online platforms to accommodate individuals with different disabilities.

One particular area that states, as well as federal agencies, have found to reach mobility-impaired individuals is through online and technology-mediated instruction and training programs. As state agencies are typically restricted in what they can offer in terms of training, they frequently collaborate with higher education institutions, namely community colleges, to offer job training for the disabled. These programs are uniquely situated to reach those with an inability to regularly attend a physical class, and through specialized software, they can be made accessible to the visually impaired. In addition, training programs such as these have been tailored to meet specific job training needs, and the result has been the promotion of interagency collaboration (Wilson, Getzel, & Brown, 2000).

Higher education institutions have also benefitted from the migration to online instruction as this capacity has allowed colleges to reach previously underserved disabled populations. This represents a shift in institutional thinking about traditional college students and expands opportunities for institutions to create partnerships with public agencies, nonprofit bodies, private corporations, and other higher education institutions. Examples include partnering with local libraries to allow computer access for mobility-impaired individuals to take online classes, seamless web-based instructional programs with community colleges that are physically adapted to those with visual limitations to complete degree programs, customized training for local industry partnerships, and even tailoring academic programs to job opportunities that might be particularly keyed to individuals with certain ability limitations. An added advantage for many institutions is also the ability to use state and federal program monies to fund the tuition of some of these individuals who would otherwise go unserved.

Online course and program offering to individuals with disabilities does impact the instructional distribution of courses, including the use of certain colors, font choice and size, voice reader capabilities, and the use of integrated media, such as video and audio embedded material. These challenges are presented to instructional designers who must find a way to engage the individual learner, with or without disabilities, and to convey content in meaningful ways. For some institutions, these challenges are made even more difficult when the coursework is targeted at specialized populations with unique needs, such as veterans. A large number of physical disabilities are and have been attributed to serving in a military capacity, and for many veterans, their perceptions of the world around them have changed in meaningful ways that can impact how they see something as basic as course content and web design.

Ultimately, online course delivery presents an innovative way of helping those with disabilities be trained and educated to compliment the workforce and help achieve individual satisfaction and growth. With a great deal of public financing available for such programs, higher education institutions have begun to realize an opportunity to fill a void in the educational market. Many of these activities, however, have been on a small scale; have been locally focused responses to a targeted need, such as a collaboration of businesses forming a workforce cooperative; and have not been expanded to serve state or district-wide needs. Online educational opportunities have the flexibility to meet diverse learner needs and have been increasingly the focus of state-level partnerships to help different populations join the workforce, and these policy imperatives will certainly continue to grow as technology becomes more interactive.

Policy Management

Policy management conceptually relates to how leaders and bureaucrats implement the rules and regulations of a particular mandate, policy directive, interpretation, or piece of legislation. These individuals can be agency directors, accountants, program officers, and even college and university trustees and presidents. In the case of online learning and the increased offering of college courses and programs of study in online and hybrid formats, this means that multiple levels of individuals can interpret and make decisions on the appropriateness of offering programs in this manner. Additionally, the discretion policy leaders have in interpreting decisions can be critical elements in how much and how many programs can be offered (Lipsky, 2010).

A clear example of how a state in the U.S. considers online learning has to do with the verification of educational providers, be they nonprofit or for-profit providers. For-profit institutions have realized a dramatic increase in enrollment, from under 750,000 in 2001 to over 2.5 million college students in 2010. During that same period of time, there has been a 225%

increase in the number of institutions offering for-profit online instruction with an average associate's degree cost of near $35,000 USD. Nearly all of these students (96%) take out loans to finance their education, and these trends have resulted in several states enacting legislation to require these institutions to have a physical presence in their state in order for students to be eligible for financial aid. The policy discretion of such an activity is the specificity of such a physical office: Must the office be a place with a classroom? Can the space be out of an individual's home? Can a rented, physical post-office box meet the standard? Such questions become problematic to answer and place such rule and regulation development in the hands of those in the policy management realm.

Policy management at the state level also requires the formation of some sort of plan, whether a formal planning document, process, or a de facto planning response that is not specific but is the result of actions and intentions. Such efforts are often based on political dispositions and beliefs, and in the case of higher education, this is typically the result of beliefs expressed in the form of nonspecific policy development—meaning actions such as legislative funding, governing board appointments, and rule and regulation creation for online learning, which all fall within a certain set of beliefs but are expressed individually through the actions rather than an expressed policy briefing development that states an official stance on a topic (Anderson, 1975).

In higher education, online learning has grown out of a dual focus on college and university growth and prestige, as well as a mounting desire to see higher education as a competitive commodity to be vetted through competition rather than coordination. The unintended result has become that public institutions increasingly focus on their prestige factor, with even community colleges building residence halls and focusing on transfer education programs rather than building job-training programs. The vocational and occupational programs as a result have moved to more of the for-profit providers, with many of these offered in online formats with higher tuition pricing and a reliance on loans for students. The net effect is that public education is less focused on practical applications of learning, and access to higher education, be it for creating social mobility or just for job training, has become the role of for-profit online providers.

CRITICAL DISCUSSION FOR POLICY LEADERS

Technology is not static, and as it changes, it can change the culture of society generally and learning specifically. Technology has integrated everything from the auto-driving experience to grocery shopping, and educational providers need to explore and continually ask how it can and will impact both the delivery of education and the expectation of students.

Through technology, learning styles have changed and evolved, and teaching styles have similarly changed and evolved. The result is a fluid state of education today, and online learning is a significant part of that fluidity.

Online learning is not the full extent of technology's impact on education, as even the basic audio-visual experience of 40 years ago has been modified to make use of new technologies. Online learning has, however, changed the education experience to one in which content is "produced" as much as it is taught. The effective online class experience is filled with multiple media episodes, including audio, visual, and interactive real-time conversations, and much of this development is aided by professionals who claim the title "instructional designer." Increasingly, these instructional designers are professionals who have a sense of television production quality video development, and as the classes are enhanced, the experience is somehow taken from the full purview of the faculty member and situated in the institution. In the U.S., this situation is the proper interpretation of intellectual property, as classes belong to the institution, but it does prompt further the question of education versus entertainment.

Beyond the technicalities of the campus experience, there are regional and national policymakers who have a role in defining the scope of online and technologically mediated college offerings. These can be state agency directors, elected council members or state legislators, and even nationally elected officials. There are at least five distinct areas that these policymakers must consider as online college instruction grows.

1. *Content rigor.* Policymakers must be aware of the need to assure the quality of a course, and although this is not limited to online instruction, it certainly includes the growing number of private, proprietary providers who use adjunct instructors to deliver course content. Policymakers must find ways, through certification or validation with some criteria, for assuring that students who pay for online instruction are indeed provided with meaningful and appropriate learning opportunities.
2. *Instructor quality.* Regional accrediting bodies in the U.S. have historically provided academic personnel qualification regulations, and these types of regulations are important for several reasons. First, they assure the consumer, in this case, college students, that the person delivering education has a level of content knowledge to appropriately teach at the postsecondary level. Second, regulation assures content can be tied to appropriate licensures. Third, instructor quality to some extent provides a proxy for course quality, meaning that if the instructor is well qualified in a content area, then presumably the course will present material at an appropriate level. And fourth, instructor quality helps to assure degree program

quality, assuming that instructional faculty have the discretion to design online courses that have a depth of content.

3. *Coordination.* Policy leaders must use the changes in how education can be distributed to coordinate efforts between public agencies for workforce and economic development, human services training, and even in adult basic education programs. With the disappearance of geographic distances, online education can be provided from geographically diverse locations, using physically separated expertise to provide education programs to those with an inability to physically relocate for education or training. Many public programs focus on specific content areas, such as a specific occupational training program. Through distance education, this training can be distributed widely with minimal cost. Policy leaders, however, must be the ones to coordinate these programs to reduce or eliminate the duplication that so frequently besets public agencies.
4. *Capacity building in higher education.* Higher education as a gateway to an enhanced quality of life has traditionally been limited to those with an ability to physically access a campus or its curriculum. With movements to online instruction, this barrier is removed to the extent that a citizenry has access to the internet, but policy leaders working in higher education can also use online programs to build access to higher education, broadly distribute high need programs a reduced cost, and even create a stronger assurance of quality across providers. Additionally, faculty expertise can be better utilized across systems, and individual classes can be structured around the best faculty member in a university system rather than just a campus.
5. *Role definition.* Perhaps the most critical discussion for policy leaders is the identification of the role of online education programs, answering the question: why is it in the best interest of the state/institution/province to offer this program or set of courses in an online format? Providers must have reasons for what they are doing, and the more that these reasons are tied to the public good and the mission of higher education, the better they can be defended in the public eye.

Directions for Future Research

Future research in online education needs to address multiple elements of the policy dimensions of regulating this segment of the higher education industry. Research should address not only how legislative and political bodies set the agenda for online education, but which bureaucrats have access to the rule and regulation implementation, and how these street-level individuals access and differentiate policy. Further, research should explore the regulation of hybrid programs and how a physical presence differentiates the educational experience, both in terms of student learning and instructional delivery. These discussions will most certainly introduce the element of the nonprofit status of many educational providers, and research must explore the effectiveness of programs in meeting learning outcomes and the role of profitability in organizational structure.

Limitations of the Discussion

This chapter discussion was limited to a particular time in history, one in which much of the online educational provision is only starting to emerge. Online learning, although over 15 years old, has gone through a rapid growth phase to one of more maturity, meaning that many of the growing pains associated with rapid expansion have begun to pass, and discussions about the meaning of offering programs online have begun to emerge. The novelty of online learning is still somewhat present, but fading, and with it, the policy discussion of its regulation will continue to evolve.

Overall, higher education institutions and leaders often do not see themselves as part of a larger system, and with an attention to serving students on their campuses, they often do not need to have this perspective. However, when seen from the larger perspective of serving a state or province, institutions have to look at how they reach out to students, and as they see the benefits and challenges of moving classes into an online format, they see that they are not alone. This means that institutions have to talk to each other to assess what others are doing, how they are providing education, and what resources can be maximized across institutions and even public sectors to best serve their public interests.

Systems behaviors can benefit the public, students, institutional mission definition and work, and ultimately, the stakeholders who make use of higher education. Online learning has matured greatly in the past decade, but it is still in its infancy in terms of integration to the larger perspective of higher education as a public service, and the next two or three decades could rewrite how colleges and universities provide a postsecondary education.

REFERENCES

Allen, E. I., & Seaman, J. (2010). *Class differences: Online education in the United States.* Neeham, MA: Sloan Center for Online Education.

Allen, E. I., & Seaman, J. (2015). *Grade change: Tracking online education in the United States.* Needham, MA: Babson Survey Research Group and Quahog Research Group.

Anderson, J. E. (1975). *Public policy-making*. New York, NY: Praeger.

Astin, A. W. (1984). Student involvement: A development theory for higher education. *Journal of College Student Personnel, 25*(4), 297–308.

Bowen, W. M., Schwartz, M., & Camp, L. (2014). *End of academic freedom*. Charlotte, NC: Information Age.

Burke, J. C. (1998). Performance funding indicators: Concerns, values, and models for state colleges and universities. In J. C. Burke & A. M. Serban (Eds.), *Performance funding for public higher education: Fad or trend?* New Directions for Institutional Research, No. 97 (pp. 49–60). San Francisco, CA: Wiley.

Chapman, B. F., & Henderson, R. G. (2010). E-learning quality assurance: A perspective of business teacher educators and distance learning coordinators. *Delta Pi Epsilon Journal, 52*(1), 16–31.

Chatterjee, L., & Mitra, M. (1998). Evolution of federal and state policies for persons with disability in the United States: Efficiency and welfare impacts. *Annals of Regional Science, 32*, 347–365.

Cody, A. (2014, June 30). Common Core-aligned tests and the new Pearson GED: Failure by design? *Education Week Teacher*. Retrieved from blogs.edweek.org/teachers/living-in-dialogue/2014/06/common_core-aligned_tests_and_.html

Dougherty, K. J., Natow, R. S., & Vega, B. E. (2012). Popular but unstable: Explaining why state performance funding systems in the United States often do not persist. *Teachers College Record, 114*(3), 1–41.

Dykman, C. A., & Davis, C. K. (2008). Online education forum: Part two-teaching online versus teaching conventionally. *Journal of Information Systems Education, 19*(2), 157–164.

Ginn, M. H., & Hammond, A. (2012). Online education in public affairs. *Online Journal of Distance Learning Administration, 15*(2). Retrieved from www.westga.edu/~distance/ojdla/summer152/ginn_hammond152.pdf

Heller, D. E. (2004). State oversight of higher education. In R. G. Ehrenbert (Ed.), *Governing academia* (pp. 49–67). Ithaca, NY: Cornell University.

Hersh, R. H., & Merrow, J. (2006). *Declining by degrees higher education at risk*. New York, NY: St. Martin's Griffin.

Holmwood, J. (2015, March). *The changing uses of the university: Neo-liberalism and the values of public higher education*. Presentation at the Annual Meeting of the Universities Association of Lifelong Learning, Durham University, United Kingdom.

Karabell, Z. (1998). *What's college for?* New York, NY: Basic Books.

Layne, M., Boston, W. E., & Ice, P. (2013). A longitudinal study of online learners: Shoppers, swirlers, stoppers, and succeeders as a function of demographic characteristics. *Online Journal of Distance Learning Administration, 15*(2). Retrieved from www.westga/~distance/ojdla/summer162/layne_bost_ice162.html

Lipsky, M. (2010). *Street-level democracy: Dilemmas of the individual in public services*. New York, NY: Russel Sage Foundation.

Marshak, L., van Wieren, T., Raeke, F., Swiss, L., & Dugan, C. (2010). Exploring barriers to college students of disability services and accommodations. *Journal of Postsecondary Education and Disability, 22*, 151–165.

Miller, M. T., & Morris, A. A. (2008). Profile of online programs in private colleges: From college to university with a click. *Academic Leadership, 6*(1). Retrieved from www.academicleadership.org

Mintz, S. (2014). The future of higher education. *Inside Higher Ed.* Retrieved from https://www.insidehighered.com/blogs/higher-ed-beta/future-higher-education

National Council for State Authorization Reciprocity Agreements (NC-SARA). (2015). Key attributes of SARA. Washington, DC: Author. Retrieved from nc-sara.org/about/key-attributes/sara

National Health Service Corps. (1992). *NHSC report 18, financial responsibility: Sound management and sound institutions.* Washington, DC: Author.

Patterson, B., & McFadden, C. (2009). Attrition in online and campus degree programs. *Online Journal of Distance Learning Administration, 12*(2). Retrieved from www.westga.edu/~distance/ojdla/summer122/patterson112.html

Olesen-Tracey, K. (2010). Leading online learning initiatives in adult education. *Journal of Adult Education, 39*(2), 36–39.

Selingo, J. J. (2013). The new, nonlinear path through college. *The Chronicle of Higher Education.* Retrieved from http://chronicle.com/article/The-New-Lifelong-Nonlinear/141867/

Sibley, W. A. (1998). *University management 2010.* Stillwater, OK: New Forums Press.

Taylor, R. E. (1984). *The unfinished agenda: The role of vocational education in the high school.* Columbus, OH: National Center for Research in Vocational Education, U.S. Department of Education.

Topper, A. (2007). Are they the same? Comparing the instructional quality of online and face-to-face graduate education courses. *Assessment and Evaluation in Higher Education, 32*(6), 681–691.

United States Department of Education. (2012, February 13). *Overview: The federal role in education.* Washington, DC: Author. Retrieved from www2.ed.gov/about/overview/fed/role.html

Wilson, K., Getzel, E., & Brown, T. (2000). Enhancing the post-secondary climate for students with disabilities. *Journal of Vocational Rehabilitation, 14,* 37–50.

CHAPTER 11

IS THE ONLINE PATH REALLY AN OFF-ROAD ADVENTURE?

The Realities of Online Education

Vance Johnson Lewis
The University of Central Arkansas

ABSTRACT

Discussions of online education often only include the positives—the access and flexibility—of the platform; however, there are many issues with online education that are seemingly ignored until after an online program has been designed and/or selected. In this chapter, the realities of online education are offered in the contexts of student involvement, campus resources, and degree viability. When considered in this context, and in combination with the facts surrounding online education, the platform's issues with student learning, resources, and public perception work together to tell a story that online education is not the pathway to an education that many claim it to be.

For many, the idea of being able to earn a college degree from the comfort of their own home holds massive appeal. Online education is so appealing that in the fall of 2011, 6.7 million students were enrolled in at least one

Technological Innovation Networks, pages 231–242

online course (Allen & Seaman, 2013). Online education, however, is not an easy path to a successful career. By examining the problems with student learning, the problems with resources, and the problems with perceptions, one can see that online education is a path fraught with pitfalls.

This discussion of online learning discusses three areas: student involvement, campus resources, and degree viability. If online learning is an acceptable alternative to the traditional classroom, then it should produce equal outcomes to traditional learning and not place students at a disadvantage either in learning or future employment. By examining the problems associated with online education, more informed decisions can be made regarding the use of online education, and more meaningful discussions can be held regarding the future of online education and online research.

Though with today's technology many classes feature some online component, certain elements need to be present for a course or degree to be considered "online." For a course to be online, at least 80% of the course materials need to be delivered via the Internet; otherwise, it is considered a traditional course at a "brick and mortar" institution (Allen & Seaman, 2006). Other factors that determine if a course is online are that the course features computer-mediated interaction, is "independent of a location," and is "time-flexible" (Bejerano, 2008, p. 409). As much of the current literature makes no distinction between online courses and degrees offered by traditional colleges and those offered by proprietary online colleges, there is no distinction made in the following discussion.

Closer examination of online education and its inherent problems is significant to higher education because of the perceived benefits of technology in all aspects of society, especially in matters of education. One need only watch television for a few minutes to hear the testimonials of professionals who advanced their careers with their online degree or how the pajama-clad young lady got her career started by going to school online. "Historians of education recount a steady cycle of failed reforms whether based on new technology, new management and budget procedures, new reward structures, or the new magic of competition" (Spence, 2001, p. 12). Online education embodies many of the qualities of this warning; thus, this discussion is important because it looks past the hype that often surrounds online education to reveal that online education has some grave drawbacks.

PROBLEMS WITHIN STUDENT LEARNING

Regardless of the personal motivation, pursuit of a degree can generally be surmised as the student's attempt to better him/herself. While an online degree can be seen as a means to this betterment, the logistics of online education can actually work against the student. Online degrees raise questions regarding the student's social interactions, academic honesty, and suitability to online learning.

For online degrees to be effective for the student, the student must have a great deal of self-motivation and organizational skills (Hale, 2007). "The biggest con [to online education] is that there tends to be a much higher dropout rate than a residential college... some [online programs] can have dropout rates of 70 percent, where 30 percent would be a high rate for a traditional campus" (Frey, 2009, p. 41). Because these online students do not have the face-to-face accountability created in the traditional classroom, many online students simply find it hard to develop their own patterns for studying and schedules for learning (Haigh, 2007).

Online students also miss out on the various campus and classroom activities that connect them to the learning process (Bejerano, 2008). The social learning theory, as related to education, discusses how the relationships between the three elements of development—environmental, cognitive, and behavioral—work together to provide one with the ability to learn (Akers, 1998). In viewing online learning in the context of the social learning theory, problems are found in the interactions students have with the educational process because "learning and knowing are shaped by the kinds of interactions a student has with others" (Hill, Song, & West, 2009, p. 89). Because the structure of online classes limits, though does not preclude entirely, the ability of students to have in-depth discussions along with the benefits of class culture and community, the online learner does not have the chance to develop to the fullest their mastery of the subject (Hill et al., 2009). The ideas presented by the social learning theory are echoed by students who credited the lack of emotional connection to other students as a major barrier in holding online conversations and seeking class solutions (Kim, Lui, & Bonk, 2005).

Though today's technology does allow for and enable students to have substantial interactions with others and their instructors, there are barriers that are still difficult for students to overcome. Although some research suggests that with proper pedagogy, online classes can be as effective as traditional courses (Driscoll, Jicha, Hunt, Tichavsky, & Thompson, 2012), predominately online students often must battle not only the lack of emotional connection but also the problems of not getting real-time feedback from their instructors and peers. Unlike in the classroom or even in the traditional office setting where immediate interactions between student and teacher can take place, the online student must wait for their questions to be answered when the instructor or peer receives it (Kim et al., 2005). Because of this timing problem, students either have to wait and therefore take longer with assignments or proceed without assistance, which may cause them to learn the material incorrectly.

The issue of academic dishonesty is viewed as a large problem for online education. Unlike their traditional counterparts, online students have the opportunity to enlist outside assistance, either in the form of a person or materials, during examinations (Sileo & Sileo, 2008). Revisiting the social learning theory, as discussed by Michaels and Meithe, though cheating may not be an intention of the online student, as the student discovers how other students in the course take advantage of outside assistance, "an attitude in which academic dishonesty is viewed as normal behavior" may arise therefore negatively impacting the actual learning within the online program (as cited in Grijalva, Nowell, & Kerkvliet, 2006, p. 181).

The quality of learning for the student may also be problematic in online learning because the student is mismatched to the online learning environment. "Students in online courses need to be self-disciplined in order to succeed" (Bejerano, 2008, p. 412). Even if it is a minimal motivation, the regularly scheduled class time and physically present classmates do present a certain level of accountability to the traditional student that that the online student does not have. The ability to postpone when they will attend the online class to another time can inherently cause the student to never attend the online class.

The online student's comprehension ability might also be problematic to the online education. As defined in the second level of Bloom's Taxonomy, comprehension is the ability of the student to interpret "words into graphics, [interpret] the meaning of certain facts, or [make] predictions for the future" (Bender, 2003, pp. 18–19). Comprehension most often takes place through discussion, which can either be face to face or through online means (Bender, 2003); however, if the student lacks keen ability for the solitary intake and process of information, then without the oral and visual components of learning, the student cannot receive the full benefits of the educational experience (Levine, 2002).

PROBLEMS WITHIN RESOURCES

Just like traditional education, online educational programs need adequate support from the higher educational institution. In many ways, online education requires more resources and investment to ensure its success. Infrastructure and technology as well as faculty commitment are essential to online programs and without these elements, online education can suffer.

Perhaps the biggest problem to a quality online education is the physical infrastructure of the program itself. For a viable online program to exist, there must be ample Internet connections as well as all of the necessary cameras and computers to broadcast the teacher and connect the students to a central online classroom (Tomei, 2002). If there is not an ample infrastructure, the volume of students logging into the course management system can cause it to run at ineffectively slow speeds or prohibit students from participating at all in the discussions or from uploading assignments by a deadline (Lyons, 2004).

Problems with the technology itself can also be damaging to the quality of an online degree. Closely tied to the problems of the infrastructure, the failure of the network or mainframe can cause lengthy delays and disruptions for the students as well as an overall frustration with the program (Lyons, 2004). When schools fail to utilize not just the correct amount of technology but also the correct types of technology, the educational experience can be more damaging to the students than educational.

Problems with technology extend from the educational site to the individual student's work station. Often, students will have insufficient and incompatible technology to support their participation in the online classroom, thus causing them to miss out on certain aspects of the educational experience (Haber & Mills, 2008). Faculty also express concern over the student's own technical knowledge and that the student may not know how the use the technology well enough to function in the online class (Haber & Mills, 2008). Because of these technological issues on the student end, it is sometimes difficult to gauge how much of the educational material the student has actually been able to study and, thus, how much of the overall education they have actually achieved.

The lack of appropriate support for the use of online education is particularly problematic to the delivery of education. Having appropriate technology supervisors on site is crucial to upholding the quality of online education. Institutions "need to provide continuous technical support for faculty members when they are developing and delivering online courses" (Tallent-Runnels, Lan, Fryer, Thomas, Cooper, & Wang, 2005, p. 168). Online education programs require round-the-clock technical supervision so that the educational experience is not disrupted for the students.

Interestingly, for traditional higher education institutions, there has been a decrease in the spending for online learning while there has been an increase in the offering of online education (Haber & Mills, 2008). Converting to the use of online education can be a costly and time-consuming undertaking, especially when done correctly, and many higher education institutions do so in the hopes that the online education programs will decrease their overall spending and increase revenue (Shedletsky & Aitken, 2001). On average, an online course costs an institution three times more than a traditional course and often, the financial backing is simply not present to uphold the quality of the online course (Shedletsky & Aitken, 2001).

Students often choose to become online students because they lack access to a traditional, onsite educational facility. A problem with many online education programs is the lack of resources available to the students.

"Institutions must make available to online students the same academic resources available to on-campus learners" (Gaide, 2004, p. 8). Without these resources, the online education is lacking in quality because students have not been able to take advantage of all the available materials for their study.

Perhaps the biggest problem regarding the quality of online technology and resources stems from the faculty themselves. Teaching an online course often takes a great deal more energy and time on the part of the teacher than a traditional course and often without the respect that a traditional course brings. In a national survey of 109 institutions, teaching an online class received almost no recognition in the tenure and promotion of faculty (Adams, 2003). For faculty who desire tenure, devoting efforts to teaching online may be minimal and therefore the quality of the instruction may be lacking.

The faculty may also be a problem for the quality of online education in the amount of time it takes to teach a quality online course. Whether it be a new subject or one with which a faculty member has experience, "the time it takes to develop a course [online] is often more than [a faculty] can devote to it" ("Encouraging Faculty," 2003, p. 5). Faculty "may be untrained and unfamiliar with the tools and technology available, making their job of translating and developing sound classroom practices" difficult to attain (Bejerano, 2008, p. 412). Even if the faculty is familiar with technology, without "continuous technical support," faculty are often reluctant and unable to create a quality online course (Tallent-Runnels et al., 2005, p. 168).

PROBLEMS IN PERCEPTIONS

It is commonly known that certain traditional higher education institutions are better regarded than others and that holding a degree from one college over another can ensure a candidate a job or graduate admission. When thinking about online programs, however, many people seem oblivious to the fact that the online degree may not be respected. In both academic and the public sectors, questions regarding the validity of online education can place a candidate at a disadvantage. Perhaps the most damaging blow to the field of online learning comes from academia itself. Although the inclusion of partial or totally online learning has continued to rise within academia, embracing those who have earned an online degree is seemingly absent within higher education. Translating an online degree into either graduate school admission or higher education employment continues to be a problem for those who have earned an online degree.

Though many online degrees are touted as a way to better one's life while not having to adjust one's work or personal schedule, online degrees may not afford their holders equal opportunity of use. Some students quickly find that the credits they earned online will not transfer to a traditional institution (Carnevale, 2002). Similarly, students who earned some or all of their degrees online have difficulties when attempting to enter graduate schools. Ironically, much of the resistance to online education comes from academia itself, with some 66% of professors stating that learning outcomes for online courses are inferior to traditional classes (Allen & Seaman, 2012). Researchers Defleur and Adams conducted a study of graduate admissions officers to see, all other things equal, if holding an online degree was viewed as a negative in the admissions process. From this study, 96% of admissions officers stated that they would admit the candidates with traditional degrees without reservations while "only 7% of the public institutions and just over 11% of their private counterparts indicated that they would recommend" admitting the applicant with a degree earned totally online (DeFleur & Adams, 2004, p. 157). Applicants with credits earned both online and traditionally received about 43% support rate from the graduate admissions officers (DeFleur & Adams, 2004). Whether from an online institution or a traditional institution with online offerings, degrees earned online face an uphill battle when looked upon as a means for graduate study.

There is also a conflict in not only how online degrees affect graduate school admissions but also who is teaching in the graduate programs. In a study conducted by Adams and DeFleur (2005), an overwhelming 98% of faculty search committee chairs responded that they would select a candidate who held a traditional graduate degree over a candidate who held an online graduate degree. Of this population, 85% stated that they had concerns about the quality of the degree that was earned online (Adams & DeFleur, 2005).

Though the issue of the quality of education continues to be a force in higher education, the issue of experience is seemingly more important to finding a job within higher education. Respondents to the Adams and DeFleur study commented that "there can be no substitute for face-to-face interaction and experiential learning" and "[r]ecommendations are an important gauge to intellectual promise and collegiality. If there were no interaction of depth how could such a judgment be measured?" (2005, p. 80). Faculty members are reluctant to hire new professors who have earned their degrees solely and even partially online because they themselves question the validity of a degree that has been earned with little to no actual interactions.

Faculty hiring decisions also question the personal quality of a candidate who decides to earn their degree online rather than in the traditional setting. Some see that the educational experience online gives graduate students and their advisors alike too much opportunity for dishonesty within the process (Adams & DeFleur, 2005). One respondent to the Adams and DeFleur study stated, "I don't believe graduate courses can be taught online. University life is part of personal education which is needed to be a social individual. Online education is misused by individuals who don't understand the difference between an educated human being and a barbarian with skills" (2005, p. 81). Being able to grow as a person is viewed as essential to the hiring practices of higher education, and this calls into question how they could view these students as unworthy in their own eyes yet actively work to graduate these students into nonacademic careers.

Employers outside of academia also express concerns about those who possess an online degree. As studied by Adams and DeFleur, hiring managers in eleven sectors of business were asked to choose between two candidates: one with a traditional degree and one with an online degree. An overwhelming 96% of the managers chose the applicant with the traditional degree with only 4% selecting the candidate who held the online degree (Adams & DeFleur, 2006). The respondents gave various reasons for these selections, with the majority falling into concerns about the lack of interaction within the education, actual skills gained in the education, and experience gained in the education (Adams & DeFleur, 2006).

Employers also express concern over online degrees because of the opportunity for fraud. Vicky Phillips, founder of GetEducated.com, states, "There are now more fake online MBA programs in the U.S. than real ones.... It's an area that's ripe for diploma mills" (Gangemi, 2005, n.p.). Because of these concerns over the legitimacy of online degrees, simply holding an online degree is not enough to gain better employment. As expressed by Gloria Odogbili, a recruiter of UBS Investment Bank in New York, "The only way we'd hire [an MBA with an online degree] is if their resume is strong and they can explain why they had to get their MBA online" (Gangemi, 2005, n.p.). Even if the online program is accredited, it may not be enough to overcome the perceptions of online learning (Carnevale, 2007).

DISCUSSION OF ISSUES

No research and discussion on the problems of online education can be truly complete with acknowledging the findings of the benefits of online education. Researchers Berker and Horn (2003) argue that the benefit of online education is found in the increased access it allows for individuals who otherwise could not pursue higher education. Many argue that the structure of online learning actually requires greater interactions between teachers and students than traditional education (Deal, 2002). Some researchers even claim research results that indicate there is no significant different in the learning outcomes of online learners to their traditional counterparts (Joy & Garcia, 2000). Given the increase in the enrollment in and pursuit of online education, being fully informed about both the problems and positives of online education is imperative for higher education.

The increased access to higher education dominates the literature on the benefits of online education. The problematic issue with this benefit is that the end result of the access may not be equal to that of students in traditional education. As stated, both employers and higher education itself question online education and do not hold it at the same value as traditional degrees. This discrepancy brings up an issue of ethics on the part of higher education as to whether online programs should be touted as an equal alternative to a traditional education. Knowing that online degrees may not be as advantageous to the student while continuing to offer and promote such degrees asks the question of whether higher education really wants to develop the student and the educated population or whether they are just in it for the increased enrollments that online education can bring.

For-profit colleges might also need to acknowledge issues of customer satisfaction in the use of their online degrees. In the spring of 2008, as HD DVD lost its battle with Blue-Ray, Best Buy Corporation offered a $50.00 gift card to anyone who had purchased an HD DVD player in an attempt to soothe the feelings of those who held the expensive, and now outdated, technology (Lawler, 2008). In 2015, Corinthian College, a former for-profit university, was found to have purposely misled students regarding the usability of their degrees and was ordered to pay damages to its former students (Kamenetz, 2015; Mitchell, 2015). Just as in the retail market, as online colleges and programs promote more of a customer service model, they might also be pushing higher education closer to a point where students—customers—might start requesting refunds when their degrees do not work and are quickly outdated.

The issue of validity and respect for online degrees certainly needs to be explored in more detail. Very little research has been done that examines the perceptions in the business world regarding online degrees. There also is certainly a need for online colleges, particularly those considered to be for-profit, to create better partnerships with the business industry so that their degrees can be respected by the business world and not placed at an automatic disadvantage to traditional degrees.

Online programs can certainly do more to establish stronger foundations for their students, especially in the area of internships. Having internships during college assists students in their social and professional skills as well as increasing their chances of obtaining relevant employment upon graduation (Knouse, Tanner, & Harris 1999). Though some online colleges and programs, such as the University of Phoenix, do require internships, this is not the industry rule. Online programs need to find ways to assist their students in obtaining internships in their geographic areas, which can assist in their subsequent job searches, but also in alleviating some of the concerns surrounding online degrees regarding social development.

Finally, the problems associated with online learning maybe indicative of a bigger picture problem. Distance education is not a new concept. Correspondence courses have been around for centuries, and multimedia elements of distance education have been present for decades. When the Internet was introduced into the equation in the early 1990s, it was heralded as the next big change in education and that traditional education was doomed (Larreamendy-Joerns & Leinhardt, 2006); however, over 20 years later, little has actually changed. Perhaps, rather than attempting to make online education the main learning tool, it has revealed itself as incapable of standing on its own, and attempting to utilize online as the main component of learning cannot and will not work.

CONCLUSION

Online education, whether viewed as a positive or a negative, is and will continue to be a part of the education landscape. Online education does allow people who otherwise could never achieve higher education a chance to reach their goals. That chance, though, may not be sufficient, and understanding the problems associated with online learning is essential to providing the best higher education possibilities.

ACKNOWLEDGMENTS

The author wishes to thank Dr. Denni Blum, Oklahoma State University; Dr. Shabana Mir, American Islamic College; and Ms. Kelsey Johnston Knuckles, Texas State University, for their invaluable input into the writing of this article.

REFERENCES

Adams, J. (2003). Assessing faculty performance for merit: An academic accomplishment index. *Journalism and Mass Communications Educator, 58*(3), 240–250.

Adams, J., & DeFleur, M. (2005). The acceptability of a doctoral degree earned online as a credential for obtaining a faculty position. *American Journal of Distance Education, 19*(2), 71–85.

Adams, J., & DeFleur, M. (2006). The acceptability of online degrees earned as a credential for obtaining employment. *Communication Education, 55*(1), 32–45.

Akers, R. (1998). *Social learning and social structure: A general theory of crime and deviance.* Boston, MA: Northeastern University Press.

Allen, I. E., & Seaman, J. (2006). *Making the grade: Online education in the United States.* Needham, MA: Sloan Consortium.

Allen, I. E., & Seaman, J. (2012). *Conflicted: Faculty and online education.* Babson Park, MA: Babson Survey Research Group.

Allen, I. E., & Seaman, J. (2013). *Changing course: Ten years of tracking online education in the United States.* Babson Park, MA: Babson Survey Research Group.

Bejerano, A. (2008). The genesis and evolution of online degree programs: Who are they for and what have we lost along the way?. *Communication Education, 57*(3), 408–414.

Berker, A., & Horn, L. (2003). *Work first, study second: Adult undergraduates who combine employment and postsecondary enrollment* (Report No. NCES-2003-167). Washington, DC: National Center for Educational Statistics. Retrieved from http://nces.ed.gov/ pubsearch/pubsinfo.asp?pubid=2003167

Bender, T. (2003). *Discussion based online teaching to enhance student learning: Theory, practice and assessment.* Sterling, VA: Stylus.

Carnevale, D. (2002). Missed connections. *Chronicle of Higher Education, 49*(8), A35.

DeFleur, M. & Adams, J. (2004). Acceptability of online degrees as criteria for admission to graduate programs. *Journal of Computing in Higher Education, 16*(1), 150–163.

Deal III, W. (2002). Distance learning: Teaching technology online. *Technology Teacher, 61*(8), 21.

Driscoll, A., Jicha, K., Hunt, A. N., Tichavsky, L., & Thompson, G. (2012). Can online courses deliver in-class results?: A comparison of student performance and satisfaction in an online versus a face-to-face introductory sociology course. *Teaching Sociology, 40*(4), 312–331.

Encouraging faculty to develop online courses. (2003). *Distance Education Report, 7*(18), 5.

Frey, C. (2009). Different paths to a college degree. *U.S. News & World Report, 146*(8), 40–42.

Gaide, S. (2004). Best practices for helping students complete online degree programs. *Distance Education Report, 8*(20), 8.

Gangemi, J. (2005). Do online MBA's make the grade?. *BusinessWeek Online.* Retrieved from http://www.businessweek.com/bschools...4192_bs001.htm

Grijalva, T., Nowell, C., & Kerkvliet, J. (2006). Academic honesty and online courses. *College Student Journal, 40*(1), 180–185.

Haber, J., & Mills, M. (2008). Perceptions of barriers concerning effective online teaching and policies: Florida Community College faculty. *Community College Journal of Research & Practice, 32*(4–6), 266–283.

Haigh, M. (2007). Divided by a common degree program? Profiling online and face-to-face information science students. *Education for Information, 25*(2), 93–110.

Hale, S. (2007). Being online. *Academe, 93*(6), 28–32.

Hill, J., Song, L., & West, R. (2009). Social learning theory and web-based learning environments: A review of research and discussion of implications. *American Journal of Distance Education, 23*(2), 88–103.

Joy, E., & Garcia, F. (2000, June). Measuring learning effectiveness: A new look at no-significant-difference findings. *Journal of Asynchronous Learning Networks, 4*(1). Retrieved from http://onlinelearningconsortium.org/jaln_article/

measuring-learning-effectiveness-a-new-look-at-no-significant-difference-findings-2/

Kamenetz, A. (2015). Corinthian Colleges misled students on job placement, investigation finds. *NPRED.* Retrieved from http://www.npr.org/sections/ed/2015/11/17/456367152/corinthian-misled-students-on-job-placement-investigation-finds

Kim, K., Liu, S., & Bonk, C. (2005). Online MBA students' perceptions of online learning: Benefits, challenges, and suggestions. *Internet and Higher Education, 8*(4), 335–344.

Knouse, S., Tanner, J., & Harris, E. (1999). The relation of college internships, college performance, and subsequent job opportunity. *Journal of Employment Counseling, 36*(1), 35.

Larreamendy-Joerns, J., & Leinhardt, G. (2006). Going the distance with online education. *Review of Educational Research, 76*(4), 567–605.

Lawler, R. (2008). Best Buy offers HD DVD owners $10 million in gift cards, trade in, shoulder to cry on. *Engadget.* Retrieved from http://www.engadget.com/2008/03/19/best-buy-offers-hd-dvd-owners-50-gift-card-trade-in-shoulder/

Levine, M. (2002). *A mind at a time.* New York, NY: Simon and Schuster.

Lyons, J. (2004). Teaching U.S. history online: Problems and prospects. *History Teacher, 37*(4), 447–456.

Mitchell, J. (2015). Corinthian colleges ordered to pay damages to students. *The Wall Street Journal.* Retrieved from http://www.wsj.com/articles/corinthian-colleges-ordered-to-pay-damages-to-students-1446059379

Shedletsky, L., & Aitken, J. (2001). The paradoxes of online academic work. *Communication Education, 50*(3), 206.

Sileo, J., & Sileo, T. (2008). Academic dishonesty and online classes: A rural education perspective. *Rural Special Education Quarterly, 27*(1/2), 55–60.

Spence, L. (2001). The case against teaching. *Change, 33*(6), 11-19.

Tallent-Runnels, M., Lan, W., Fryer, W., Thomas, J., Cooper, S., & Wang, K. (2005). The relationship between problems with technology and graduate students' evaluations of online teaching. *Internet and Higher Education, 8*(2), 167–174.

Tomei, L. (2002). *The technology façade: Overcoming barriers to effective instructional technology.* Boston, MA: Allyn & Bacon.

CHAPTER 12

TECHNOLOGICAL INNOVATION IN HEALTH PROFESSIONS EDUCATION

Meeting the Needs of a 21st Century Workforce through Enhanced, Blended and Online Learning

Paige L. McDonald, Elizabeth A. Ruckert, Ozgur Ekmekci, Jacqueline S. Barnett, and Howard Straker
The George Washington University School of Medicine and Health Sciences

ABSTRACT

In this chapter, we discuss three essential competencies for future healthcare professionals and our use of technology as a "bridge" for attaining these competencies. Specifically, based on specific pedagogical goals, VoiceThread, iPads, Panapto, Google Drive, and Blackboard Collaborate were used to help students develop higher order thinking skills and make collaborative, evidence-based decisions. Though the integration of pedagogy and technology has allowed us to promote competency in higher-order thinking, collaborative

Technological Innovation Networks, pages 243–270

problem solving, and evidence-based decision-making, further consideration must be given to addressing student preferences for face-to-face interaction and instructor-centric learning environments. We also argue that institutions must provide technologies that marry with the pedagogical aims of programs of study. In addition, sufficient instructional design expertise is required to assist with program and course design and implementation of technology.

Institutions of higher education face challenges in educating a 21st-century workforce. There is an increasing demand to develop competencies graduates will need to survive in an ever-changing, dynamic workforce, while continuing to impart requisite content knowledge and skills. Health professions education, in particular, faces the challenge of imparting the knowledge and skills necessary for graduates to readily participate in interdisciplinary teams that ensure high-quality care in increasingly complex healthcare environments. Meeting this challenge requires enhancing existing curricula to include active and collaborative learning strategies that develop higher-order thinking, collaborative problem solving, and evidence-based decision-making skills. Yet these types of curricula enhancements can prove quite challenging for educational programs already charged with meeting accreditation standards, increasing class size, and decreasing the cost of health professions education. Technology can be used to address the challenges while capturing the benefits of these curricular enhancements.

Integration of technology into traditionally face-to-face education models as enhanced, blended, or fully online course offerings allows for increased class size and more efficient use of face-to-face class time. In addition, it can be implemented without adding time or content to the already saturated curricula, thereby not impacting accreditation or increasing cost. Innovative uses of technology can facilitate adoption of the active and collaborative learning models necessary to prepare future health professionals for the workforce. In this chapter, we provide examples of how our selection of technology for enhanced, blended, and online learning in health professions education at the George Washington University (GW) supports pedagogical methods essential to the development of competence in higher-order thinking, collaborative problem solving, and evidence-based practice.

PROBLEM AND CONTEXT

At the turn of the 20th century, when infectious diseases were the primary cause of death and life expectancy from birth in the U.S. was only 47.3 years (U.S. Department of Health and Human Services, 2000), the first generation of health education reform was born. Led by the 1910 Flexner Report (Duffy, 2011; Flexner, 2002), early reform efforts focused on establishing the biomedical model of training and implementing educational curricula grounded in science and clinical experiential learning. These reform efforts led to advances in medicine and science and equipped physicians and health professionals with the requisite knowledge and skills to combat infectious disease and extend life expectancy (Frenk et al., 2010) to 76.7 years by the end of the 20th century (Arias, 2002).

In direct contrast to the 1900s, the current U.S. healthcare system is struggling to care for an increasingly aging population, in which nearly fifty percent of adults suffer from at least one chronic disease (Ward, Schiller, & Goodman, 2014). In fact, chronic diseases have replaced infectious diseases as the leading cause of death and disability (Centers for Disease Control and Prevention, 2015). Furthermore, health inequities, medical errors, behavioral and environmental risks, exorbitant cost, and an inadequate understanding and response to the changing population demographics continue to threaten the health of the nation.

While the nation's populations and health concerns have changed over the past century, health professional education has not adequately kept pace with new pedagogies or adequately responded to new challenges (Frenk et al., 2010; Interprofessional Education Collaborative Expert Panel, 2011; Nancarrow, Moran, & Graham, 2014; Pruitt & Epping-Jordan, 2005). Many health professionals are being educated using static and outdated curricula that lack adequate training in teamwork and interprofessional care, the use of technology in the delivery of care (Frenk et al., 2010), patient- and population-oriented care, and care models that emphasize continuous and primary care over acute and specialty care (Frenk et al., 2010; Pruitt & Epping-Jordan, 2005).

To equip health professionals with the requisite knowledge and skills to confront the multifaceted concerns of the current healthcare system, educational institutions can no longer rely on 20th-century educational pedagogies and philosophies. A new era of health education reform is needed to promote transformative learning and interdependence (Frenk et al., 2010), which will facilitate the ability of graduates to collaborate in interprofessional teams, deliver coordinated patient-centered care, and apply evidence-based approaches to improve the quality of care and health outcomes.

Health professions education is considered competency-based education (CBE), as learners are expected to meet certain observable and measurable outcomes in knowledge, skills, and attitudes (Gruppen, Mangrulkar, & Kolars, 2012). These "competencies" define the work domains of a particular health discipline as performed in the context of providing care (Interprofessional Education Collaborative Expert Panel, 2011). Although many health professions have revised their professional competencies to better align with the needs of the current healthcare system (Accreditation Council for Graduate Medical Education, 2014; American Academy of Physician Assistants, 2012; Federation of State Boards of Physical Therapy,

2006; National Organization of Nurse Practitioner Faculties, 2014), the ability of educational programs to incorporate them into the curricula remains a challenge, rendering graduates inadequately prepared to deliver high-quality care in a challenging healthcare system. Future health professionals will need to generate and apply new knowledge, engage in critical thinking and decision making, and utilize and adapt professional competencies in a collaborative manner to deliver system-based care for specific patient situations.

In efforts to improve the 21st-century healthcare system, the Institute of Medicine (IOM) released sentinel reports on safety, quality, and healthcare disparities (Institute of Medicine, 1999, 2001, 2002). Following these reports, the IOM promoted educational reform to better align health profession education with the needs of the 21st-century healthcare system and recommended that all health professionals establish the following core competencies for future practice: provide patient-centered care, work in interdisciplinary teams, use evidence-based practices, apply quality improvement, and utilize informatics (Institute of Medicine, 2003).

To better prepare health professionals to meet the needs of patients in a technologically advanced and complex healthcare system, we recommend all graduates achieve the following three competencies for 21st century practice: higher-order thinking, collaborative problem solving, and the ability to apply evidence in making clinical practice decisions. Although we might use different terminology than the IOM, these three competencies are inferred from and congruent with the IOM recommendations and also serve as requisite foundational skills to achieving the IOM competencies. Higher-order thinking, collaborative problem solving, and the ability to apply evidence to make clinical decisions will equip future health professionals with vital skills to answer the IOM's call to translate knowledge into clinical practice, work in high-performing efficient teams, and use technology safely and effectively to improve care delivery and transform the healthcare system (Institute of Medicine, 2003). Ultimately, the three competencies outlined in this chapter will support the "Triple Aim" of healthcare reform, which is to improve the patient's experience of care, improve population health, and reduce healthcare costs (Institute for Healthcare Improvement, 2015; Mahon, Henderson, & Kirch, 2013).

Our methods for achieving these competencies are supported by various pedagogical approaches to learning and consider potential challenges and future recommendations for educating health professionals. We realize that adopting new pedagogical approaches may require innovation in delivery models. Traditional delivery models require physical presence in classrooms, placing boundaries on enrollment numbers and creating challenges for innovative delivery and collaborative learning (Ruckert et al., 2014) and, therefore, can inhibit competency development. Online education (technology-enhanced, blended, and fully online courses), when combined with innovative pedagogies, can facilitate competency development and higher levels of learning (McDonald, Straker, Schlumpf, & Plack, 2014; Ruckert et al., 2014). Our examples highlight how we have aligned pedagogical methods with online delivery models to develop specific competencies.

PEDAGOGICAL MODELS TO ASSIST IN COMPETENCY DEVELOPMENT: EXPERIENTIAL, ACTIVE, AND COLLABORATIVE LEARNING

As described above, the knowledge, skills, and attitudes needed to be a competent health professional in today's complex healthcare environment require achievement of high-level learning objectives. Helping learners to achieve deeper levels of critical thinking and processes for problem solving is important not only in the classroom but for future professional success. Learners must be equipped with the ability to continue directing their own learning long after their time in the classroom (be it physical or virtual classroom) is done (National Research Council, 2000). For example, the ethical standards of the physical therapy profession include language related to lifelong learning (American Physical Therapy Association, 2011). The Core Values of Excellence for physical therapist offers "engaging in acquisition of new knowledge throughout one's professional career" as an indicator of professional excellence (American Physical Therapy Association, 2011, p. 71). Next we review the educational literature informing the learning events we designed to develop these skills in our learners.

Experiential learning theory serves as the foundation for active and collaborative learning models that use learners' prior experiences to move beyond rote memorization of information toward deeper meaning, such as application, synthesis, evaluation, and new knowledge generation (Dewey, 1938; Garrison, Anderson, & Archer, 2003; Kolb, 1984). Dewey and, later, Kolb asserted that learning occurs when learners compare current experiences to prior conceptualizations, make meaning (or generate new knowledge) from the comparison, and determine how the new conceptualization might be useful for future action. A dialogic relationship between action and reflection enables learners to make meaning of experiences through a process of pre-reflection, reflection, and post-reflection (Dewey, 1938; Kolb, 1984). Kolb developed a cycle of four abilities learners require to engage actively in the process of learning: (a) be able to fully engage and participate in an experience, (b) be able to critically reflect upon the experience in relation to prior experience or perspectives, (c) be able to formulate observations from reflections into new conceptualizations, and (d) be able to actively experiment with new conceptualization or apply them in new situations (Kolb, 1984).

Health professions educators applying Kolb's model must be able to create classroom environments that allow students to cycle between active engagement with new concepts, reflection upon that engagement, and application of knowledge generated in new ways. Incorporation of active learning strategies provides this opportunity in the classroom. Active learning techniques, no matter the duration, help learners take a step back and think about the content at a deeper level (Chickering & Gamson, 1991; National Research Council, 2000; Silberman & Auerbach, 2006). Active learning strategies have long been considered an essential component of "good" teaching, helping students move away from memorization and toward active engagement with the material through discussion, debate, presentations, and reflection (Chickering & Samson, 1991, 1999). Common active learning strategies include think-pair-share, the one-minute paper, concept mapping, fishbowl, and polling (Silberman & Auerbach, 2006). Indeed, active learning helps to create meaningful experiences for a learner to interact with and create new personal meaning and new knowledge (Dewey, 1938; Kolb, 1984).

Collaborative learning theory describes how individuals make meaning of prior experiences through social interaction with peers and instructors (Garrison et al., 2003; Garrison & Kanuka, 2004; So & Bush, 2008): "an individual person constructs his or her knowledge through the process of negotiating meaning with others... cognitive development is highly dependent on social interaction and collaboration with more capable and knowledgeable others" (So & Bush, p. 3). The community of inquiry (COI) framework provides guidance on how instructors can facilitate higher levels of learning through a process of collaborative meaning construction (Garrison et al., 2003; Garrison & Kanuka, 2004). Application of this model requires ensuring cognitive presence, social presence, and teaching presence in the learning environment (Garrison et al., 2003; Garrison & Kanuka, 2004). Establishing a sense of community among learners ensures the sense of safety and cohesion required to critically evaluate and debate controversial topics, which promotes higher levels of learning and enables collaborative problem solving (Garrison et al., 2003; Garrison & Kanuka, 2004; McDonald, Straker, et al., 2014; Ruckert et al., 2014). Applying the framework also requires facilitated opportunities for pre-reflection, reflection, and post reflection essential to the meaning-making process (Garrison et al., 2003). Future health professionals will be required to critically evaluate complex health situations and engage with others in providing solutions to complex health problems. Facilitating communities of inquiry within health professions education supports achievement of this competency.

The collective importance of each learning theory described above is integrated into Figure 12.1. The key concepts of experience, reflection, active engagement, and collaborative learning are emphasized. Each of the instructional design exemplars shared later in this chapter includes elements of these learning theories. In addition, our exemplars cross the health professions' "classroom continuum" from technology-enhanced face-to-face instruction to blended to fully online instruction. Table 12.1 presents definitions we use to differentiate the degree of technology integration in each of these delivery models.

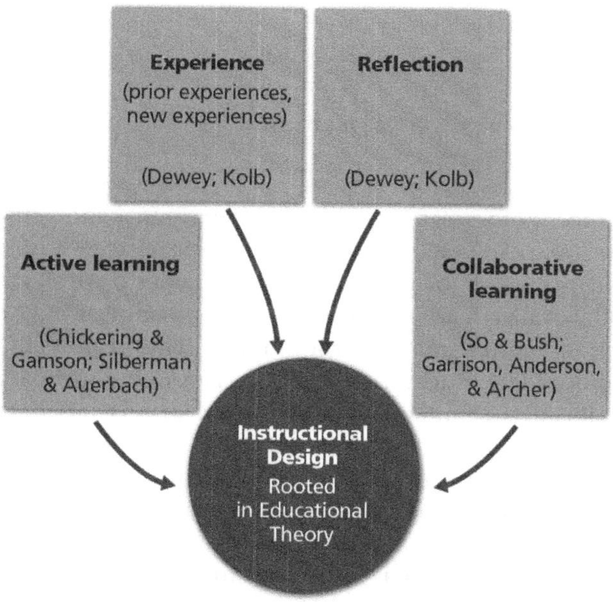

Figure 12.1 Instructional design rooted in educational theory.

TABLE 12.1 Learning Delivery Model Definitions

Delivery Model	Definition
Technology-enhanced learning	*Supplementing traditional* face-to-face courses with online technology for a specific pedagogical purpose without reducing seat time (Allen & Seaman, 2014; McDonald, Lyons, Straker, Barnett, Schlumpf, & Corcoran, 2014)
Blended learning	*Integrating online* learning and face-to-face learning based upon pedagogical goals while reducing seat time (McDonald, Lyons, et al., 2014; Picciano, 2009)
Online learning	Adopting *fully online* courses with or without the requirement for synchronous interaction among students (McDonald, Lyons, et al., 2014)

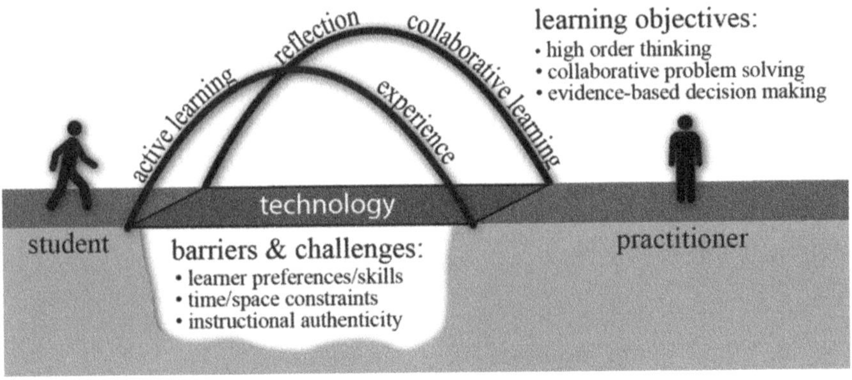

Figure 12.2 Technology as a bridge for barriers and challenges to achieving health practitioner competencies.

Adopting technology can promote the experiential and active learning processes necessary to help individuals learn best regardless of face-to-face or online environments (National Research Council, 2000). It can also support the collaborative learning required to promote meaning negotiation, higher levels of learning, and critical thinking (Garrison et al., 2003; Garrison & Kanuka, 2004; Ruckert et al., 2014). However, technology must be chosen specifically and integrated and aligned with clearly defined pedagogical goals (Picciano, 2009). The subsequent exemplars demonstrate our alignment of technology-enhanced, blended, and online delivery with pedagogical models to build competency in higher-order thinking, collaborative problem solving, and evidence-based decision-making. In each exemplar, technology served as a "bridge" for us to address challenges to competency development. Figure 12.2 presents how we integrated educational theory and technology to facilitate learners moving from student to practitioner with these competencies.

COMPETENCY 1: HEALTH PROFESSIONALS MUST DEMONSTRATE HIGHER-ORDER THINKING

Conceptual Definition

Higher-order thinking, which is firmly grounded in the progressive learning model originally developed by Bloom, Engelhart, Furst, Hill, and Krathwohl (1956) and later modified by Anderson and Krathwohl (2001), is generally accepted to be a set of cognitive abilities that allows individuals to move beyond simply understanding basic knowledge in a given content area to synthesizing information—with the ultimate outcome of being able to evaluate content in that domain (Collins, 2014). This type of thinking not only enables people to solve existing problems, but it also allows them to define new problems and to apply knowledge gained in problem solving to new and differing situations (Brookhart, 2010).

In today's world, where individuals have to utilize multiple information sources to identify and solve complex problems (by constructing, evaluating, and discerning from among available scenarios and solutions), employing higher-order thinking has become a vital necessity for successful outcomes to be achieved (Tilchin, 2015). This necessity is nowhere more visible than it is in today's healthcare industry, at a time when all stakeholders are scrambling to find solutions that reduce cost and improve quality, while an aging population is presenting many new challenges on both the problem and solution sides of the equation (Brennan & Sullivan-Marx, 2012).

Main Challenges Associated With Naturally Demonstrating Higher Order Thinking

In response to the need to prepare future clinicians to solve complex problems, educational programs aimed at preparing healthcare professionals for successful clinical practice have started incorporating a greater number of activities that develop and promote higher-order thinking skills in their curricula (Prober & Heath, 2012). Despite this trend, there are still significant challenges associated with successful adoption and implementation among students in the classroom. Our undergraduate and graduate offerings at GW are no exception.

First of all, we observe that—while our students are generally skilled in using technology for social purposes (mainly broadcasting or forwarding information)—they are not always inclined to use it to generate new knowledge or demonstrate higher-order thinking. Some of this may be attributed to learning habits that have formed throughout their traditional educational experiences (Brookhart, 2010), whereby they have been asked to primarily repeat memorized knowledge as opposed to identifying and creatively solving problems. Settling into this learning pattern limits students to remaining within the confines of stated problems (instead of identifying root causes) and prescribed solutions (instead of developing innovative solutions).

Students' lack of tolerance of ambiguity (Furnham & Ribchester, 1995) serves as a second challenge. "Real-world" application of knowledge, especially in healthcare settings, involves situations that do not always have the structure and facts available leading to one "correct" answer. Often, healthcare scenarios involve weighing pros and cons and determining the "best" answer based on patient preferences, clinical experience, and evidence-based practice. Many students struggle with scenarios that do not have black-and-white "answers." They perceive such learning scenarios as stressful and undesirable, and they may "shut down" their learning as a result. Low tolerance for ambiguity has been found in medical students, regardless of their year of training (Weissenstein, Ligges, Brouwer, Marschall, & Friederichs, 2014), which likely also applies to other health professions students. Technology also provides easy access to a wealth of information, causing students to falsely think they can quickly find the answer to a problem. We have found that our students often accept information found on the internet as fact without verifying the accuracy or credibility of the source. Student bias for efficiency and technology's ease of access to information prevents some students from exploring alternative information surrounding a problem or potentially "gray" area. They look to technology to help them quickly find "the" answer without considering other alternatives to a problem, which limits or prevents the occurrence of higher order thinking.

A third challenge arises when students are asked to support their coursework with reference material other than those provided in class. In creating such support, students primarily rely on web technology to retrieve third-party information readily available through singular sources (as opposed to evaluating multiple sources, synthesizing relevant information, and creating their own added value). This reliance on singular sources significantly limits the boundaries of creative thinking and reduces diversity of opinion. Even if students try to move up to higher levels of learning in Bloom's taxonomy (1956), the direction they take is often subject to potential bias in the work from which they have heavily borrowed when supporting their own ideas.

The fourth significant challenge we observe in our classrooms, which affects our students' ability to naturally integrate higher order thinking into their coursework, is that traditional learning assessment methods remain fairly limited in their ability to evaluate higher-order thinking. Multiple choice questions, which are likely the most widely used format of learning evaluation in health professions education (Downing & Yudkowsky, 2009), remain limited in measuring higher-order thinking, especially when evaluating collaborative problem solving in teams (Valentine, Nembhard, & Edmondson, 2015). Creating examination questions that assess these levels of learning is more challenging. Unless we create proper alignment among learning objectives, classroom activities, and assessment methods (all of which should reside at the upper echelon of Bloom's taxonomy), it is unrealistic to set up a learning environment in which students can be expected to naturally demonstrate higher-order thinking.

An Example of the Use of Technology to Promote Higher-Order Thinking

A major priority for our faculty is to encourage students to use technology to supplement and support their classroom activities—whether these are weekly readings or assignments—with information they find on their own. While faculty teaching our classes do not specify any particular resource for this purpose, given the era in which we live, students rely heavily on the web in seeking such resources. Our faculty—aware of the fact that easy access to an abundance of information often makes it all too tempting for students to rely on singular and (usually non-peer-reviewed) sources—clearly point out in course syllabi that a minimum of five peer-reviewed sources should be used to support assignments.

To encourage students to access, evaluate, synthesize, and critique multiple sources, one of the assignments used in the capstone course taken at the end of our online master's programs involves working in small groups of five to six students. A virtual role-play activity is used to set up a scenario in which the groups are given a case study from the Harvard Business School Press depicting problems faced by an executive leader at a healthcare organization. The groups are asked to work as a consulting entity to analyze the case and to prepare a 10-slide VoiceThread presentation to be delivered to the executive leader featured in the case study. "VoiceThread© is a cloud application into which instructors and students can upload documents, images, presentations, audio files and videos to share with either private groups or the entire community" (Ruckert et al., 2014, p. 62). Students are also asked to integrate at least five peer-reviewed sources into their presentation.

Once the presentations are finalized within each group, the groups are then paired with one another to review and evaluate the other group's presentations in asynchronous online discussion forums facilitated by the instructor, which becomes part of students' assignment grades. This assignment is designed to allow students the opportunity to collaboratively: (a) analyze a given case to identify organizational problems; (b) create a set of solutions to be recommended to the organization, while synthesizing additional sources of information into their presentation to create new knowledge; and (c) evaluate similar (new) knowledge created by their peers that is grounded in the same objective facts depicted in a common case study (which illustrates the potential diversity of thought that culminates through the exchange of various perspectives).

In the same capstone course, our faculty members use asynchronous online discussion forums to also set up individual peer-review assignments to promote higher-order thinking by dividing the class roster into triads (i.e., groups of three students), whereby each student in a triad is asked to review the draft proposals (these could be quantitative research proposals, organizational change proposals, or quality improvement proposals, depending on the course they are taking) developed by the other two members of the triad and to provide constructive feedback that will help their peers strengthen their proposals. This activity is set up using role-play, where each student is asked to play the part of a foundation president who has to evaluate proposals submitted for funding and respond with a one-page decision letter that supports key reasons as to why the proposal was either accepted or rejected. This asynchronous online activity forces students to analyze and evaluate their classmates' proposals and create added value for them within a coherent logical structure (as opposed to simply sharing superficial comments, such as "great job!!!" or "loved your proposal!" without creating any added value.)

In each of the examples provided, collaborative case resolution and peer review of proposals facilitates the higher-order thinking (analysis, evaluation and knowledge generation) that can result from active learning strategies and COIs. Analysis, application, and evaluation (cognitive presence) are facilitated as students work in groups to solve problems, critique the solutions of other groups, and analyze the proposals of peers. Students also have multiple opportunities to reflect upon action (their group resolutions or individual proposals): first, as peers provide critiques of their work, and second, as they consider how they will integrate that feedback in future revisions of their proposals. Student-to-student interactions in class discussions prior to group work or peer review triads promotes the social presence required to allow thoughtful evaluation and critique of peer work. Faculty presence and feedback also proves critical to guiding the critique of case resolutions and peer proposals toward more in depth analysis and evaluation.

Future Recommendations

In addition to using role-play-based asynchronous online discussion forums and VoiceThread to facilitate small group peer reviews and presentations, web-based simulations involving more complex scenarios in which students receive dynamic information and are asked to make decisions (based on the information they receive) can be a very effective and rewarding learning experience for students. By employing such simulations, both as teaching and evaluation platforms, student learning experiences in the classroom can move closer to the actual organizational or clinical environment in which they will be asked to make similar decisions and take similar actions. As most successful decisions in life require higher order thinking, attaining mastery of this competency in class is crucial for success in the organizational and clinical settings.

While simulations have been shown to be very effective in helping students develop higher-order thinking in classroom and clinical settings, cost-effective web-based simulations involving healthcare scenarios are still limited. At this time, the cost of integrating simulations into each course remains high, competing against the need to lower the cost of health professions education. Another issue with using asynchronous technology platforms in which students are asked to perform group-based activities is the challenge associated with evaluating competencies at the individual level, while also taking into account the quality of the collaborative effort. Robust (and fairly time-consuming) evaluation approaches are required to assess each student, as well as the group itself, as it pertains to demonstrating competency in terms of higher-order thinking. Research is also required on models of assessment facilitating individual and collective contributions are required.

COMPETENCY 2: SKILL IN COLLABORATIVE PROBLEM SOLVING

Conceptual Definition

Future healthcare professionals will require skill in collaborative problem solving in interdisciplinary teams to solve the complex health issues of an ever-expanding and aging population (Brennan & Sullivan-Marx, 2012; Reinke & Hammer, 2011). With regard to education, collaborative problem solving is an active learning strategy that can be used to enhance critical thinking (Gokhale, 1995; Ramsden, 1992). Exploring challenging problems in small groups of students supports self-learning (Boud & Feletti, 1991) and promotes critical thinking (Yin, 2011), skills learners require for negotiating future real-world complexities. Correspondingly, health professions curricula must incorporate this active learning strategy to promote higher-order thinking skills (Prober & Heath, 2012) and train students to solve problems collectively.

Main Challenges to Promoting Competence in Collaborative Problem Solving

Educators hoping to adopt collaborative learning models, which require active engagement in the learning process, face numerous challenges. Class size presents a significant challenge. Dividing large sections of students into the smaller groups required for collaboration and active engagement in the process of learning becomes challenging in large classroom settings. At GWU, we have experienced this challenge in our graduate courses comprised of over 60 students who meet in either small classrooms or lecture halls that are not conducive to small-group activities.

In addition to the environmental or space barriers to collaborative problem solving, the traditional lecture-based models of health education delivery perpetuate a reliance on a more passive role in the learning process. As a result, faculty adopting more active learning models may face resistance from students who have only experienced learning as a process of memorization of lecture material prepared by a subject-matter expert rather than as a process of collaborative knowledge generation with peers and instructors (McDonald, Straker, et al., 2014). At GW, when blending a first-semester physician assistant (PA) course, all class lectures were placed online (OL). Though students actively participated in face-to-face (FTF) activities and assessments indicated higher levels of learning with this model, student feedback indicated a preference for the "sage on the stage" model of delivery in which the instructor presented the material or gave them the "answer" to controversial issues.

An Example of the Use of Technology to Promote Collaborative Problem Solving

Health, Justice & Society (HSJ) is a second-semester course in the physician assistant program at GWU that introduces students to social determinants of health and health disparities. The course confronts controversial topics such as bias and ethics in healthcare, challenging students to consider how their own biases might influence future practice and to collectively resolve ethical dilemmas similar to those they may face in practice. Achieving these objectives necessitates students' active participation in the process of learning to critically evaluate, synthesize, and ultimately apply course concepts to controversial real-world healthcare situations. Moreover, resolution of ethical cases requires negotiation of the meanings of dilemmas presented in relation to future practice with their peers.

In 2013, we redesigned HSJ from a traditional face-to-face course into a blended course and have offered the course in this format to three successive cohorts of students. Currently, this is the only blended course offered in our PA curriculum. The blended course model allows us to create a learning environment in which learners can collaborate with peers, collectively negotiate solutions to complex problems, critically reflect upon a variety of differing solutions, and discuss how they will apply what they have learned in future clinical practice. The course requires weekly online (OL) activities and biweekly face-to-face (FTF) sessions. This structure allows us to maximize student–student and student–faculty interaction in both learning environments, which is critical to adopting collaborative inquiry as a pedagogical approach.

Applying the COI framework (Garrison et al., 2003; Garrison & Kanuka, 2004) necessitated creating smaller learning groups of six to eight students among cohorts that have consistently exceeded 60 students. Assigning students to smaller learning groups for the duration of the course supports development of the sense of "community" required to safely explore biases and negotiate the meaning of controversial course topics. Figure 12.3 presents a model of our learning environment.

In our class, knowledge was developed individually, in small groups, and as a larger community of inquiry over the course of the semester related to issues topics like bias, social determinants of health and ethics. Requiring students to post individual responses to discussion questions and assigned cases prior to collaborating in their groups (in both OL and FTF discussions) ensures individual cognitive presence. Teaching presence within the OL discussions and in FTF discussions ensures individual and group contributions that are aligned with intended learning outcomes. Toward the end of the course, when students are more comfortable in their learning groups, they are tasked with solving four ethics cases by applying knowledge gained during the semester in relation to future clinical practice.

Technology proves critical to facilitating collaboration and the negotiation of meaning in this activity. We use VoiceThread technology for this activity because it invites interactivity in the form of comments on virtual presentations. Prior to preparing for group collaboration on the ethics

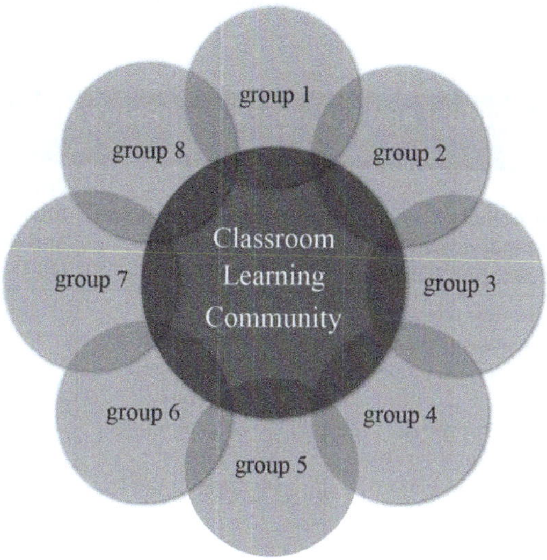

Figure 12.3 Learning community in health, justice, and society.

cases, students view a VoiceThread presentation on ethics developed by a faculty member (allowing familiarity with the technology), read required materials, and take a quiz on ethics to ensure foundational understanding of content. Then, students post individual recommendations for case resolution in their group discussion boards. Students must post their own initial responses before they are able to view the responses of other members of their group. Once they have reviewed each other's postings, groups are given both OL and FTF time to discuss their collective resolution and create a VoiceThread presentation of that resolution. Each VoiceThread presentation is then virtually "presented" to the entire class. Following the virtual presentations, each student is required to comment on the presentation of at least two other groups, either discussing differences in comparative resolutions or posing questions for the group to consider further. These comments facilitate further negotiation of meaning among the larger learning community. Finally, students write individual reflection journals regarding what they learned throughout the collaborative learning process and from comparing their individual and group solutions to the solutions of others in the learning community. Table 12.2 details the sequencing of OL and FTF activities.

These cases are especially complex and rely on the sharing of multiple viewpoints and experiences to develop potential "solutions." The sharing of information from individual → group → class helps students develop deeper personal and intersubjective meanings related to the content. It also encourages recognition of the value of different perspectives in problem exploration and resolution, which will be critical to solving future problems in clinical teams.

TABLE 12.2 Individual and Collaborative Inquiry Related to Ethics in PA Curriculum

Assignment	Pre-Activity and Individual Application (OL)	Collaborative Meaning Negotiation and Application (OL and FTF)	Post Activity Meaning Negotiation and Reflection (OL)
Ethics cases	1. Review related OL readings and materials, including case 2. Post individual resolution in team discussion board	1. Discuss case resolution with team in OL and FTF sessions 2. Create and upload VoiceThread presentation of case	1. Review presentation of two other teams posting comments and critiques 2. Respond to questions from peers on group presentation 3. OL reflective journal entry on comparisons of resolutions

Students have responded very positively to the collaborative learning environment, specifically indicating an appreciation for learning how to collaborate and solve complex problems in groups. When asked what they learned in the course, one student noted learning about "*my thought process* in ethics, how to work in a group for tough decision making, and where my own thoughts were on topics I had yet to think about," and another remarked learning "that there are *many different perspectives* and approaches to a single problem/issue. There is never one right answer." However, not all students appreciated the requirement to use technology for case presentation, interaction, and collaboration. In particular, students do not see the point of interacting in the OL environment when they could meet FTF. A few teams also had difficulty using the presentation technology, which may have influenced their opinions on the use of VoiceThread to present their cases. Student criticism reflects a preference for FTF interaction and a lack of experience in using technology for learning. Recommendations that follow discuss these issues.

Future Recommendations

Student criticism of the use of technology in a blended course relates to their assumptions about the use of technology in healthcare settings. Many students think healthcare practice involves face-to-face interaction with patients and colleagues, and they do not believe they will have to use technology in future practice to interact with others (McDonald, Straker, et al., 2014). Even though future healthcare reform will require increased digital literacy on the part of practitioners (Ruggeri, Farrington, & Brayne, 2013), students are not yet convinced they will be required to collaborate with colleagues and patients via technology. To improve student buy-in related to technology, faculty can include a primer related to use of technology in physician assistant practice at the beginning of the semester. Explicitly presenting literature on patient interactions with telehealth, interdisciplinary conferencing across health systems in different geographic areas, or legislation related to electronic medical records practices may help to achieve student buy-in related to the use of technology in the classroom.

Our inability to adopt technology that mirrors what students will actually use in practice intensifies the challenge of student acceptance of technology in learning. For example, to promote a more authentic use of technology for problem solving, we partnered groups of students with practicing clinicians to discuss real-world manifestations of ethical dilemmas discussed in the course. This was successful in terms of providing students with clinical partners to ask questions and solve problems related to today's healthcare climate. The collaboration could have been strengthened through the use of a more sophisticated video conferencing system. We utilized Blackboard Collaborate, which students have available as an application on iPads provided by the GW PA program. However, the iPad application does not allow for video-conferencing, only audio conferencing with viewing of moderator materials, so it does not mirror the type of conferencing performed with future patients. Convincing students of the value of technology in the classroom as a way to build competency for future practice may require adoption of technology that more readily aligns with that which they will use in practice. Yet programs will need to decide how to select specific technology adoptions from the variety of technologies used in healthcare practice, particularly given the rate at which technologies change. Research is required on specific technologies that promote authentic learning.

COMPETENCY 3: APPLYING EVIDENCE IN DECISION-MAKING

Conceptual Definition

Evidence-based practice is an essential competency for today's healthcare professionals. The term evidence-based practice (EBP) describes the use of research evidence in a critical and judicious manner that results in best outcomes for a unique patient (Sackett, Rosenberg, Gray, Haynes, & Richardson, 1996). Clinicians rely not only research evidence but also on patient preferences and clinician's own clinical experience to make sound clinical decisions when practicing in an evidence-based manner (Sackett et al., 1996). EBP has since been incorporated into the practice standards of many health professionals. In 2005, Dawes and colleagues published the Sicily statement on evidence-based practice in BMC Medical Education. The report stated: "All healthcare professionals need to understand the principles of EBP, recognize EBP in action, implement evidence-based policies, and have a critical attitude to their own practice and to evidence. Without these skills, professionals and organizations will find it difficult to provide 'best practice'" (p. 1). Today's healthcare professional must be able to find evidence, appraise the quality of that evidence, and then determine how that evidence applies to the unique patient for whom he/she is caring. For example, in the physical therapy profession, Doctor of Physical Therapy (DPT) programs are required to ensure that all graduates achieve competencies related to EBP and research. Table 12.3 includes the specific objectives of the Commission on Physical Therapy Education's Evaluative Criteria for EBP (CAPTE, 2014). As described earlier in this chapter, higher-level learning is required to attain objectives involving critical evaluation, integration, application, creation, and consistency that are necessary of an

TABLE 12.3 Evidence-Based Practice: Professional Practice Expectations for Physical Therapists (CAPTE, 2014)	
CC-5.21	Consistently use information technology to access sources of information and clinical decisions.
CC-5.22	Consistently and critically evaluate sources of information related to physical therapy practice, research, and education, and apply knowledge from these sources in a scientific manner and to appropriate populations.
CC-5.23	Consistently integrate the best evidence for practice from sources of information with clinical judgment and patient/client values to determine the best care for a patient/client.

evidence-based practitioner. Therefore, EBP is a complex skill that requires ongoing didactic practice for student competence in the clinical setting.

Main Challenges to Promoting Competence in Applying Evidence in Decision-Making

A number of challenges exist for developing competence in EBP for health professions students and practicing clinicians. These challenges are noted in the literature (Manns, Norton, & Darrah, 2015; Salback, Jaglal, Korner-Bitensky, Rappolt, & Davis, 2007; Scurlock-Evans, Upton, & Upton, 2014) and have been experienced directly by the authors of this chapter as educators and clinical practitioners.

First, although there is a plethora of evidence available to students, it is challenging to determine the quality of that evidence. With the growth of internet resources, open access journals and medical library access, clinicians have more access to evidence on a given topic than ever before. Unfortunately, the "quick" search is often preferred over the intensive, targeted EBP search process. Jansen, Rasekaba, Presnell, and Holland (2012) found that general search engines such as Google were the second most commonly used way to search for answers to a clinical question, used by 83% of the sample. Even if a student does use high-quality search tools (such as PubMed and medical library resources), his or her ability to interpret the evidence, determine its applicability to a given patient, and decipher the quality of the methodology is a challenge for most students. In their systematic review of barriers toward EBP in physiotherapists, Scurlock-Evans and colleagues (2014) found that self-reported confidence in appraising methodology, psychometrics, and statistics was rated as less than 50%. Providing students with opportunities to practice appraisal skills and apply critical thinking strategies to research evidence is important in health professions education.

Time presents another challenge to implementing EBP skills in clinical practice. Finding and appraising literature related to a given clinical question takes time. Only 8% of subjects surveyed by Salback and colleagues (2007) had protected time in their standard work day to complete literature searches and critique evidence. For those without protected time, productivity standards and workload characteristics limit the amount of time available for employing EBP strategies. Literature suggests that use of EBP strategies in the first year of employment is especially low (McEvoy, Williams, Olds, Lewis, & Petkov, 2011), which is surprising given that students have recently graduated from programs emphasizing EBP. EBP often is less of a priority for therapists with limited clinical experience who are focused on trying to develop psychomotor and decision-making skills. These therapists most often look to colleagues' clinical experience and opinion before looking to the evidence to help make decisions (Manns et al., 2015).

An Example of the Use of Technology to Promote Evidence-Based Decision-Making

Technology has been effective in helping students develop evidence-based appraisal skills in anticipation of clinical practice, while also confronting some of the aforementioned challenges. A more obvious way of using technology for this purpose has been improving student awareness of professional resources available to them on the internet. Library resources, professional organization websites, and health databases are widely available to support healthcare professionals in more easily accessing evidence-based information. One example we use frequently in the physical therapy program curriculum is the Rehabilitation Measures Database (www.rehabmeasures.org), which is designed to provide summary information on reliability, validity, and other psychometric properties for a wide variety of commonly utilized tests and measures in the physical therapy literature.

An activity used in the geriatric patient management course in the physical therapy program requires students to evaluate current literature related to the management of patients with Alzheimer's disease (AD). The learning objectives for the activity are fourfold:

1. Access professional literature databases for the purpose of organizing a graduate-level paper based on evidence-based practice for the care of the geriatric client
2. Engage in peer review and development activities including critique of current literature, formulation of professional opinions for papers, evaluations, plan of care, and interventions

3. Create relevant interventions that demonstrate evidence-based, patient/client-centered practice for case studies
4. Present and debate relevant interventions with peers

The activity was based on six articles selected by the instructor that represented various aspects of physical therapy treatment for individuals with AD. An important component of the activity was use of technology, especially Google Drive, for students to demonstrate their critical appraisal skills as well as critical thinking skills. Please see Table 12.4 for a description of the activity. In addition, students were very motivated to participate in the activity because the information they were gathering and critiquing was the underpinnings of a future course activity involving students working in clinic with individuals with AD where they would have to apply what they learned. The relevance and "need to know" of the information was clear, which is important to adult learners (Merriam, Caffarella, & Baumgartner, 2007).

Use of technology had many benefits. Google Drive enabled students to provide comments to one another in real time. Specifically, some pairs of students could engage in "chat" or conversation during peer review regarding their thoughts/interpretations of a given article. Not all students provided comments to one another synchronously; most performed this asynchronously. The depth of student questions demonstrated their knowledge of evidence-based practice, which the instructor could evaluate and even provide additional comments or questions to help them "dig deeper." See Table 12.5 for sample questions and comments included in the paper demonstrating complex thinking processes. Use of PowerPoint presentations

TABLE 12.4 Description of Enhanced Course Design

	Before FTF Class	During FTF Class	After FTF Class
Activity	1. Students critique an assigned research article. 2. Students review a peer's draft of critical appraisal of a different article. 3. Students provide constructive feedback on peer's paper content.	1. Students work in small groups with the same assigned article to create a presentation for peers based on the key points of the article. 2. Students teach peers key aspects of patient care related to their article.	1. Students are encouraged to review and reflection on videos of the presentations. 2. Students apply what they learned in authentic environment (community setting) with patients living with AD.
Technology	• Blackboard • Google Drive	• Google Drive • PowerPoint	• Panapto • Electronic documentation • Reflection

TABLE 12.5 Sample Student Responses on Peer Review of AD Activity

Demonstration of EBP Appraisal Skills	Student 1: "Maybe it is worth reiterating that without a control group, treatment effect cannot truly be assessed?" Student 2: "Perhaps make it more clear in which direction the bias would exert its effect." Student 3: "Did they perform follow-up with the [subjects] later?" Student response to comment #3: "They didn't mention that… I would have loved a longer follow-up to compare their baseline/1-week post [results]."

during the in-class group work forced students to synthesize their groups' decisions about key information from their assigned article in a concise and straightforward way. Presenting this information to their colleagues helped achieve higher levels of understanding. Panapto, a lecture capture software system, was another beneficial technology that helped achieve session objectives. This software enabled all students to review and revisit the class material after the class session ended, which was helpful because each student was required to read only one article, yet had access to key concepts presented on all six articles, making the process more efficient for them. This was a way to ensure that students could review material that was presented by their peers more than once if desired.

FUTURE RECOMMENDATIONS

There are a number of recommendations for educators who are designing learning events to improve evidence-based practice skills. Manns et al. (2015) suggested the importance of helping students learn to make evidence-based decisions under time constraints to better mimic the real-world clinical environment. We agree with these recommendations. Commonly, students are assigned article critiques or clinical questions as an out-of-class activity that enables them to deliberate on and analyze content for a student-chosen amount of time. In contrast, clinical decisions in the authentic healthcare environment must be made efficiently after reviewing patient charts and discussing with the patient and/or other healthcare team members, and even in between patient sessions. Designing in-class activities that have specific and shorter time requirements for students to make evidence-based decisions may better prepare students for healthcare practice.

Evidence-based practice requires not only knowledge-based thinking skills but also communication skills to inform patients and other healthcare workers about rationale for given clinical decisions. Devising clinical scenarios in which student communication can be developed and practiced with others is a potentially helpful educational method for instructors. Specifically, patient clinical scenarios that require students to role-play the process of finding, evaluating, and delivering evidence-based education to patients and other healthcare colleagues in real time may be beneficial. Technology can help to create these role-playing opportunities in online and blended environments through synchronous or asynchronous discussions with specific time guidelines as part of the assignment instructions.

Finally, it is important for instructors to consider alternative methods of assessment that not only capture a student's article critique skills but also consider their efficiency in application of these skills and the environment in which they are assessed (Downing & Yudkowsky, 2009). One option is to include evidence-based practice decision-making activities within practical examinations that more commonly focus on psychomotor clinical skills like history taking and objective examination. Another option is to design assessments that occur in the clinical environment. In our geriatric patient management course, we follow the in-class evidence-based practice activity with an application activity with patient volunteers with AD to help students apply information in the clinical environment, but this was a relatively low-stakes activity. Having to apply these skills in a higher-stakes examination, such as a standardized patient examination, may help students translate the skills to high-pressure clinical situations. Another option is teaming up with clinical partners in the community where students are on internships. Clinical instructors could then evaluate students' EBP/appraisal performance in action in the clinical environment. Authentic learning experiences such as these, however, take considerable time and resources for faculty to develop and execute. Research on faculty workload, assessment models, and effective instructional design would provide insight into strategies to overcome these obstacles.

CONCLUSION

In this chapter, we discussed three essential competencies for future healthcare professionals and our use of technology as a bridge for attaining these competencies. Specifically, VoiceThread, iPads, Panapto, Google Drive, and Blackboard Collaborate were used to help students develop higher order thinking skills and make collaborative, evidence-based decisions. Each technology was deliberately chosen based on specific pedagogical goals, as well as the challenges we faced in the classroom (online or face-to-face). VoiceThread enabled us to address space constraints of classrooms designed for more lecture focus than small-group focus. Google Drive and VoiceThread were used to help create a collaborative learning environment that students felt safe in providing constructive feedback and asking questions to promote higher-order thinking. Use of Blackboard Collaborate to interact with practicing clinicians helped create an authentic environment for student learning by connecting students with practicing clinicians. The use of this technology also appealed to the student preferences for face-to-face interaction. We also provided training on the use of Blackboard Collaborate to develop their skill in using technology for course-related collaboration. As previously noted, Figure 12.2 models the way in which we have combined technology and pedagogy to bridge the barriers and challenges to building the competencies students require as future practitioners.

Though the integration of pedagogy and technology has allowed us to promote competency in higher-order thinking, collaborative problem solving, and evidence-based decision-making, further consideration must be given to addressing student preferences for face-to-face interaction and instructor-centric learning environments. Additionally, the selection of educational technology must be more closely aligned with the technology students will use in future practice. This requires further consideration on the part of institutions in the process of selecting potential platforms for delivery of online learning. Finally, further thought must be given to the time it requires for faculty to design authentic learning experiences that mirror the environments in which students will be practicing. These learning experiences must ultimately help students attain required competencies through didactic assessments as well as future clinical practice. In the health sciences department at GW, we have the ability to partner faculty with instructional designers to create course designs that marry appropriate technology with specific pedagogical goals. In the absence of such a partnership, faculty may find the types of instructional design processes we have described overwhelming or overly time-consuming and may opt out of attempting more innovative designs. Institutions must consider how to support faculty in the process of redesigning courses to promote competency development through incorporation of active pedagogies and appropriate use of technology.

With regard to institutional policy regarding adoption and integration of technology, institutions must provide technologies that marry with the pedagogical aims of programs of study. In addition, sufficient instructional design expertise is required to assist with program and course design and implementation of technology. Finally, institutions should provide networking opportunities for faculty to promote ongoing course improvement and integration of technology from a pedagogical approach.

DEDICATION

We dedicate this chapter to our dear colleague, Dr. Ozgur Ekmekci. Ozgur was a true academician and leader. He was committed to delivering a quality education to students and empowering and promoting faculty in their professional development. He approached life and work with the highest level of intellect and integrity. We are proud to have worked with him on this chapter. Ozgur holds a special place in our hearts, and he is truly missed as a friend and colleague. We are honored to have had the chance to know, learn from, and celebrate such a wonderful colleague.

REFERENCES

Accreditation Council for Graduate Medical Education. (2014). ACGME: Common program requirements. Retrieved from http://www.acgme.org/acgmeweb/Portals/0/PFAssets/ProgramRequirements/CPRs_07012016.pdf

Allen, I. E., & Seaman J. (2014). *Grade change: Tracking online education in the United States, 2013.* Babson Survey Research Group, Pearson and the Sloan Consortium. http://sloanconsortium.org/publications/survey/grade-change-2013

American Physical Therapy Association. (2011). *Today's physical therapists: A comprehensive review of a 21st century health care profession.* Retrieved from http://www.apta.org/uploadedFiles/APTAorg/Practice_and_Patient_Care/PR_and_Marketing/Market_to_Professionals/TodaysPhysicalTherapist.pdf

American Academy of Physician Assistants. (2012). *Competencies for the physician assistant profession.* Retrieved from https://www.aapa.org/threeColumnLanding.aspx?id=2173

Anderson, L. W., & Krathwohl, D. R. (Eds.). (2001). *A taxonomy for learning, teaching and assessing: A revision of Bloom's taxonomy of educational objectives.* New York, NY: Longman.

Arias, E. (2002). *United States life* tables, *2000. National vital statistics reports* (Vol. 51). Hyattsville, MD: National Center for Health Statistics.

Bloom, B. S. (Ed.), Engelhart, M. D., Furst, E. J., Hill, W. H., & Krathwohl, D. R. (1956). *Taxonomy of educational objectives: Cognitive domain.* New York, NY: Longmans, Green & Co.

Brennan, A. M., & Sullivan-Marx, E. (2012). The paradigm shift. *Nursing Clinics of North America, 47*(4), 455–462.

Brookhart, S. (2010). *How to assess higher order thinking skills in your classroom.* Alexandria, VA: ASCD.

Boud, D., & Feletti, G. (1991). *The challenge of problem based learning.* London, England: Kogan Page.

Commission on Accreditation in Physical Therapy Education (CAPTE). (2014). *Evaluative criteria for PT programs.* Retrieved from http://www.capteonline.org/uploadedFiles/CAPTEorg/About_CAPTE/Resources/Accreditation_Handbook/EvaluativeCriteria_PT.pdf

Centers for Disease Control and Prevention. (2015). Chronic diseases: The leading causes of death and disability in the United States. Retrieved from http://www.cdc.gov/chronicdisease/overview/index.htm

Chickering, A. W., & Gamson, Z. F. (1991). *Applying the seven principles for good practice in undergraduate education.* San Francisco, CA: Jossey-Bass.

Chickering, A. W., & Gamson, Z. F. (1999). Development and adaptations of the seven principles for good practice in undergraduate education. *New Directions for Higher Education (80)*, 75–81.

Collins, R. (2014). Skills for the 21st century: Teaching higher-order thinking. *Curriculum & Leadership Journal, 12*(14), 1.

Dawes, M., Summerskil, W., Glasziou, P., Cartabellotta, A., Martin, J., Hopayian, K., Porzsolt, F., Burls, A., & Osborne, J. (2005). Sicily statement on evidence-based practice. *BMC Medical Education. 5*(1). Retrieved from http://www.biomedcentral.com/1472-6920/5/1

Dewey, J. (1938). *Experience and education.* New York, NY: Simon & Schuster.

Downing, S., & Yudkowsky, R. (2009). *Assessment in health professions education.* New York, NY: Routledge.

Duffy, T. P. (2011). The Flexner report: 100 years later. *The Yale Journal of Biology and Medicine, 84*(3), 269–276.

Federation of State Boards of Physical Therapy. (2006). Standards of competence. Retrieved from https://www.fsbpt.org/Portals/0/Content%20Manager/PDFs/free-resources/StandardsOfCompetence2006_10.pdf

Flexner, A. (2002). Medical education in the United States and Canada. From the Carnegie Foundation for the Advancement of Teaching, Bulletin Number Four, 1910. *Bulletin of the World Health Organization, 80*(7), 594–602.

Frenk, J., Chen, L., Bhutta, Z. A., Cohen, J., Crisp, N., Evans, T., . . . Zurayk, H. (2010). Health professionals for a new century: transforming education to strengthen health systems in an interdependent world. *Lancet, 376*(9756), 1923–1958. http://doi.org/10.1016/S0140-6736(10)61854-5

Furnham A., & Ribchester T. (1995). Tolerance of ambiguity: A review of the concept, its measurement, and application. *Current Psychology, 14*(3), 179–199.

Garrison, D. R., Anderson, T., & Archer, W. (2003). A theory of critical inquiry in online distance education. In W. G. Moore, & T. Anderson (Eds.), *Handbook of distance education* (pp. 113–127). Mahwah, NJ: Erlbaum.

Garrison, D. R., & Kanuka, H. (2004). Blended learning: Uncovering its transformative potential in higher education. *Internet and Higher Education, 7*(2), 95–105.

Gruppen, L. D., Mangrulkar, R. S., & Kolars, J. C. (2012). The promise of competency-based education in the health professions for improving global health. *Human Resources for Health, 10*, 43. doi:10.1186/1478-4491-10-43

Gokhale, A. A. (1995). Collaborative learning enhances critical thinking. [Online]. Retrieved from http://bpkpenabur.or.id/paper-surabaya.htm

Institute for Healthcare Improvement. (2015). *The IHI triple aim initiative.* Retrieved from http://www.ihi.org/engage/initiatives/tripleaim/Pages/default.aspx

Institute of Medicine. (1999). *To err is human: Building a safer health system. Committee on Quality of Health Care in America* (1st ed.). Washington, DC: National Academies Press.

Institute of Medicine. (2001). *Crossing the quality chasm: A new health system for the 21st century.* Committee on Quality of Health Care in America. Washington, DC: National Academy Press.

Institute of Medicine. (2002). *Unequal Treatment: Confronting racial and ethnic disparities in health care.* Committee on Understanding and Eliminating Racial and Ethnic Disparities in Health Care. Washington, DC: National Academy Press.

Institute of Medicine. (2003). *Health professions education: A bridge to quality.* Washington DC: National Academies Press.

Interprofessional Education Collaborative Expert Panel. (2011). *Core competencies for interprofessional collaborative practice: Report of an expert panel.* Washington, DC: Interprofessional Education Collaborative.

Jansen, L., Rasekaba, T., Presnell, S., & Holland, A. E. (2012). Finding evidence to support practice in allied health: peers, experience and the internet. *Journal of Allied Health, 41,* 154–161.

Kolb, D. (1984). *Experiential learning: Experience as the source of learning and development.* Englewood Cliffs, NJ: Prentice-Hall.

Manns P. J., Norton, A. V., & Darrah J. (2015). Cross-sectional study to examine evidence-based practice skills and behaviors of physical therapy graduates: is there a knowledge-to-practice gap? *Physical Therapy, 95,* 568–578.

Mahon, K. E., Henderson, M. K., & Kirch, D. G. (2013). Selecting tomorrow's physicians: The key to the future health care workforce. *Academic Medicine, 88*(12), 1806–1811. Retrieved from http://doi.org/10.1097/ACM.0000000000000023

McDonald, P. L., Lyons, L., Straker, H., Barnett, J., Schlumpf, K., & Corcoran, M. (2014). Educational mixology: A pedagogical approach to faculty development and course redesign in Health Sciences. *Online Learning, 18*(4), 33–50.

McDonald, P. L., Straker, H., Schlumpf, K., & Plack, M. (2014). Learning partnerships: Students and faculty learning together to facilitate reflection and higher order thinking in a blended course. *Online Learning, 18*(4), 73–93.

McEvoy M. P., Williams, M. T., Olds, T. S., Lewis, L. K., & Petkov J. (2011). Evidence-based practice profiles of physiotherapists transitioning into the workforce: A study of two cohorts. *BMC Medical Education, 11,* 100.

Merriam S., Caffarella R., & Baumgartner L. (2007). *Learning in adulthood: A comprehensive guide* (3rd ed.). San Francisco, CA: Jossey-Bass.

Nancarrow, S. A., Moran, A. M., & Graham, I. (2014). Preparing a 21st century workforce: Is it time to consider clinically based, competency-based training of health practitioners? *Australian Health Review: A Publication of the Australian Hospital Association, 38*(1), 115–117. Retrieved from http://doi.org/10.1071/AH13158

National Organization of Nurse Practitioner Faculties. (2014). *Nurse Practitioner Core Competencies Content: A delineation of suggested content specific to the NP core competencies.* Retrieved from http://c.ymcdn.com/sites/nonpf.site-ym.com/resource/resmgr/Competencies/NPCoreCompsContentFinalNov20.pdf

National Research Council. (2000). *How people learn: Brain, mind, experience and school: Expanded edition.* Washington, DC: National Academies Press.

Picciano, A. G. (2009). Blending with a purpose: The multi-modal model. *Journal of Asynchronous Learning Networks, 13*(1), 1–9.

Prober, C. G., & Heath, C. (2012). Lecture halls without lectures: A proposal for medical education. *New England Journal of Medicine, 366*(18), 1657–1659. DOI:10.1056/NEJMp1202451

Pruitt, S. D., & Epping-Jordan, J. E. (2005). Preparing the 21st century global healthcare workforce. *BMJ (Clinical Research Ed.), 330*(7492), 637–639.

Ramsden, P. (1992). *Learning to teach in higher education.* London, England: Routledge.

Reinke, L. F., & Hammer, B. (2011). The role of interprofessional collaboration in creating and supporting health care reform. *American Journal of Respiratory and Critical Care, 184*(8), 863–864.

Ruckert, E., McDonald, PL., Birkmeier, M., Walker, B., Cotton, L., Lyons, L., & Plack, M. (2014). Using technology to promote active and social learning experiences in health professions education. *Online Learning, 18*(4), 51–72.

Ruggeri, K., Farrington, C., & Brayne, C. (2013). A global model for effective use and evaluation of e-learning in health. *Telemedicine and e-Health, 19*(4), 312–321. doi:10.1089/tmj.2012.0175.

Sackett, D. L., Rosenberg, W. M., Gray, J. A., Haynes, R. B., & Richardson, W. S. (1996). Evidence based medicine: what it is and what it isn't. *BMJ,* 312, 71–72.

Salback, N. M., Jaglal, S. B., Korner-Bitensky, N., Rappolt, S., & Davis, D. (2007). Practitioner and organizational barriers to evidence-based practice of physical therapists for people with stroke. *Physical Therapy, 87,* 1284–1303.

Scurlock-Evans, L., Upton, P., & Upton, D. (2014). Evidence-based practice in physiotherapy: A systematic review of barriers, enablers and interventions. *Physiotherapy,* 100, 208–219.

Silberman, M. L., & Auerbach, C. (2006). *Active training: A handbook of techniques, designs, case examples, and tips* (3rd ed.). San Francisco, CA: Pfeiffer.

So, H., & Bush, T. A. (2008). Student perceptions of collaborative learning, social presence and satisfaction in a blended learning environment: Relationships and critical factors. *Computers & Education, 51*(1), 318–336.

Tilchin, O. (2015). Computer-mediated assessment of higher order thinking development. *International Journal of Higher Education, 4*(1), 225–231.

U.S. Department of Health and Human Services. (2000). *Healthy people 2010: Understanding and improving health* (2nd ed.). Washington, DC: U.S. Government Printing Office.

Valentine, M., Nembhard, I., & Edmonsdon, A. (2015). Measuring teamwork in health care settings: A review of survey instruments. *Medical Care, 53*(4), 16–30.

Ward, B. W., Schiller, J. S., & Goodman, R. A. (2014). Multiple chronic conditions among U.S. adults: A 2012 update. *Preventing Chronic Disease, 11,* 130389. Retrieved from http://www.cdc.gov/Pcd/issues/2014/13_0389.htm

Weissenstein A., Ligges S., Brouwer B., Marschall B., & Friederichs, H. (2014). Measuring the ambiguity tolerance of medical students: A cross-sectional study from the first to sixth academic years. *BMC Family Practice, 15*(6), 1–12.

Yin, Y. K. (2011). Collaborative problem solving methods towards critical thinking. *International Education Studies, 4*(2), 58–62.

ABOUT THE CONTRIBUTORS

Inés Alegre received her PhD in management from IESE Business School and M.S. in industrial engineering from Universitat Politècnica de Catalunya, Spain. She is currently assistant professor in the faculty of economics and social sciences of the Universitat Internacional de Catalunya, and she also collaborates with other universties such as IESE and UNIR and institutions like EVPA. Her research and publications are focused on the area of entrepreneurship and social entrepreneurship. She has published several book chapters and research papers in addition to presenting her research at prestigious conferences like the Academy of Management Annual Meeting.

Jacqueline S. Barnett, DHSc, MSHS, PA-C is an adjunct faculty in the department of physician assistant studies at the George Washington University School of Medicine and Health Sciences, and an assistant professor in the department of community and family medicine, Duke University Physician Assistant Program. She received her doctorate in health sciences with a concentration in leadership and organizational behavior from A.T. Still University. Dr. Barnett's scholarly and research interests include improving learning experiences for students with disabilities, curriculum enhancement through the integration of technology and blended learning formats, interprofessional education, and improving health equity and outcomes in underserved areas.

Jasmina Berbegal-Mirabent received her PhD in management and M.S. in industrial engineering and in engineering management, all from Universitat Politècnica de Catalunya (UPC), Spain. She is currently working as

Technological Innovation Networks, pages 271–277

an associate professor at Universitat Internacional de Catalunya and as a part-time professor at UPC. She has been a visiting research associate at the Institute of Education at University College London. She is the associate editor of the *Journal of Innovation and Knowlegde* and is on the editorial board of the *Journal of Business Research and Management Decision.* Her research interests are in the areas of economics of education, technology transfer, and academic entrepreneurship.

Paolo Boccardelli is full professor of management and corporate strategy at LUISS Guido Carli University and dean of LUISS Business School (Rome, Italy). He is member of the Lars Magnus Ericsson Foundation Scientific Committee and of the EFMD Women on Business School research group. He is also designated as an expert of international standing for the Australian Research Council. A renowned speaker at international conferences (Strategic Management Society, Academy of Management, Continuous Innovation Network), he authored a large number of scholarly publications in the field of strategic management, innovation, creative industries, and business model innovation.

Daniele Bologna is a doctoral candidate in the department of psychology at the University of Cincinnati. His primary research interest focuses on personal energy management in the workplace, often conceptualized as employee engagement, job burnout, unanticipated demands/resources, or break-taking behaviors. His work has been presented at the Academy of Management and Society for Industrial and Organizational Psychology conferences and has appeared in academic publications such as the *Journal of Health Organization and Management.* He holds a Master's degree in industrial-organizational psychology from Xavier University.

Federica Brunetta is assistant professor of corporate strategy and technology and innovation management in the department of business and management of LUISS Guido Carli University in Rome. She got her PhD from the Catholic University of Sacred Heart (Italy) and spent a year as visiting scholar at the Ross School of Business, University of Michigan. Her research interests cover topics in the field of strategy and management of innovation, such as R&D management, alliances, network analysis, industry dynamics, and management of collaboration. She has authored scholarly publications and performed research projects at the national and international level.

Manuel Castriotta is a post-doctoral research fellow at the University of Cagliari (Italy). His current research regards the intersection between the fields of innovation and strategic management. Methodologically, he combines both patent technological positioning and science mapping approaches by exploiting bibliometric techniques. He has published in international journals such as *International Small Business Journal* and *Scientometrics,* as well as several book chapters and conference papers.

Maria Chiara Di Guardo is full professor of innovation management at the University of Cagliari (Italy). Her research focuses on how to organize efficiently the innovation process as well as the relationship between innovation and entrepreneurship. She has published a number of research articles for high-standing international journals, including *International Small Business Journal, Research Policy, Journal of Technology Transfer, Small Business Economics, Long Range Planning, Strategic Organization, Science and Public Policy, The Service Industries Journal, Management Research, Scientometrics,* and *Journal of Infometrics.* She is also the author or co-author of books related to innovation management and entrepreneurship.

Ozgur Ekmekci, EdD, is an associate professor of clinical research and leadership at George Washington University School of Medicine and Health Sciences. He has 25 years of experience as a manager, educator, and organizational consultant. Dr. Ekmekci serves on the review board of the Society for Advancement of Management. Dr. Ekmekci has certifications in Myers-Brigs Type Indicator and Lean Six Sigma Sensei. His research focus is organizational identification, professional identity, and interprofessional education. He is most interested in interventions in the classroom and the workplace that improve the way interprofessional teams learn and perform.

Israël Fortin studies identity and communication dynamics across levels of interaction in innovation networks and ecosystems. He holds a PhD in management from HEC Montreal and is currently a postdoctoral researcher in the department of technology and operations at the Rotterdam School of Management. His work will also appear in the *Scandinavian Journal of Management.*

Stacie Furst-Holloway is an associate professor in the department of psychology at the University of Cincinnati. Her research interests include the impact of technology on individual and team-based work behaviors as well as the implications of balancing work and home demands. Her work has appeared in the *Journal of Applied Psychology, Journal of Management, Personnel Psychology,* and the *Journal of Organizational Behavior.* Stacie frequently works as a consultant with organizations in areas of virtual work, retention, and HR assessment and evaluation. She holds a PhD in management from the University of North Carolina at Chapel Hill.

Luca Giustiniano (PhD in management/information systems) is associate professor of organization studies at LUISS Guido Carli University (Rome, Italy) and Associate Dean for Quality Management and Accreditation at

LUISS Business School. During his career he has been visiting scholar at the Viktoria Institute (Sweden, 1999), Visiting Associate Professor at the Sauder School of Business–University of British Columbia (Canada, 2011), at the ICOA (Aarhus University, Denmark, 2014) and at the Nova SBE, Lisbon (Portugal, 2015). He is a member of ASSIOA (Associazione Italiana di Organizzazione Aziendale), EGOS (European Group for Organizational Studies), and AOM (Academy of Management).

Adrián Guerrero, holds an M.Sc. in management from Universitat Internacional de Catalunya and bachelor in production engineering from Politecnico di Torino. He is currently a research assistant in the faculty of economics and social sciences of the Universitat Internacional de Catalunya. He will start his PhD in management at Universitat Internacional de Catalunya in the following months.

Trayan Kushev is an assistant professor in the management and human resources department at Cal Poly Pomona. He holds a PhD from the University of Louisville. Trayan's primary research interest lies in the area of entrepreneurship. More specifically, he studies the decision-making and cognitive processes of entrepreneurs engaging in the entrepreneurial process. Trayan's research has been published in the *Journal of Entrepreneurship*. Outside of academia, Trayan has experience working with both for-profit and nonprofit startups as well as consulting for startups.

Dr. Vance Johnson Lewis is an assistant professor of management at the University of Central Arkansas. He earned a doctorate of education in higher education and management at Oklahoma State University and completed a post-doctorate in marketing and management at the University of Florida. Dr. Lewis' primary areas of research are in organizational behavior, specifically in abusive leadership, narcissism, and job satisfaction. He also researches managerial communications and organizational training methods. He is an active member in the Academy of Management and the Midwest Academy of Management and is on the editorial board for *The Journal of Leadership and Organizational Studies.*

Na Liang is a PhD candidate in Shanxi University. Her research interests include the Republic of China era firms' human resources and the evolution of the organizational structure, as well as technology innovation. Her current research has been indexed in CSSCI.

Mei-Yan Lu is a professor of educational leadership in the Connie Lurie College of Education at San Jose State University where she previously held appointments as the associate dean of the college and as a faculty member in instructional technology. The majority of Dr. Lu's work has focused on understanding and accommodating cultural differences among secondary and postsecondary teachers and learners and how technology can be a factor in mitigating these differences. Her work has also focused on leadership issues in education, ranging from leadership training in social justice to strategies for shared governance in higher education.

Elona Marku is a PhD student in economics and business at the University of Cagliari (Italy). In the academic year 2015–2016 she was visiting scholar at Columbia Business School (USA) where she conducted research on innovation and strategy, with particular focus on patent quality and mergers and acquisitions.

Paige McDonald, EdD, is an assistant professor of clinical research and leadership at the George Washington University School of Medicine and Health Sciences and program director for the Graduate Health Sciences Core Curriculum. Her research interests focus on the scholarship of teaching and learning, in particular enhancing and blending with technology to achieve pedagogical goals. Recent peer-reviewed publications focus on collaborative learning, blended learning, reflective practice, faculty development, and technological integration in physician assistant and health professions education.

Michael T. Miller is the senior associate dean of the College of Education and Health Professions at the University of Arkansas where he also serves as professor of higher education and coordinator of the education concentration in the public policy program. His college enrolls over 1,500 online learners in programs ranging from bachelor's degrees through doctoral degrees. His research has primarily focused on the role of community in identity development as well as leadership issues in formal education settings. He has particularly worked with the issue of rural environments in educational attainment and the role of online learning.

John Mueller is an assistant professor at California State University–Fresno where he teaches entrepreneurship. His research interests include entrepreneurial finance and how new, young, and small firms handle their financial and accounting activities; technology entrepreneurship; and sports entrepreneurship. Along with his academic experience, he has spent time starting and running several businesses in the software and golf industries, with his work experiences allowing him to live and work in the United States, Europe, the Middle East, and Africa. Mueller holds a PhD in entrepreneurship from the University of Louisville, an MBA from the University of Illinois, and a BBA in management information sciences from Southern Methodist University.

Tong Quan is a lecturer at Shanxi University. She received a PhD from Nankai University, Tianjin, China. She has published a book and several academic papers, which have been indexed by CSSCI. Her major interest is in Japan's economic and social research.

Wasko Rothmann was post-doctoral researcher for media management at the European University Viadrina and is now a business consultant at an international consulting firm.

Elizabeth Ruckert, PT, DPT, NCS, GCS is an assistant professor and director of the MedStar National Rehabilitation Network and George Washington University neurologic physical therapy residency program. Her research focus is in the scholarship of teaching and learning, and fall prevention in persons with neurologic disorders. Recent peer-reviewed publications and presentations focus on the areas of blended learning, instructional technology, and instructional design in physical therapy and health sciences education. Dr. Ruckert serves on the American Board of Physical Therapy Residency and Fellowship Education Accreditation Services Committee, and is an active member of the American Physical Therapy Association.

Tao Shi is a professor and the tutor of PhD students in Shanxi University. He has published many personal books and academic papers. His major research interests include modern Chinese economic history, ancient Chinese economic history, and the dynamic of economic history.

Howard Straker, PA-C, MPH, is a physician assistant and assistant professor in the department of physician assistant studies at the George Washington University School of Medicine and Health Sciences. His research interests are in blended learning for health professions education and preparation of health professional students for practice in underserved communities. Recent peer-reviewed publications focus on collaborative learning, blended learning, reflective practice, and technological integration in physician assistant and health professions education.

Heinz-Theo Wagner is professor of management and innovation at the German Graduate School of Management and Law (GGS), Heilbronn, Germany. His research interests include innovation management, the role of digital technologies and their impact on organizations, and the IT business-value creation process.

Fei Wang is a graduate student in the master in Shanxi University. Her research interest is grain price, the market integration, and resource allocation efficiency in Qing Dynasty, especially the regionally microscopic study in Shanxi province.

Matthias Wenzel is a post-doctoral researcher at the European University Viadrina, Germany. His current research centers on strategic change, demand-side innovation strategies, digitization, and research methods.